DETECTOR OWNER'S FIELD MANUAL

by

Roy Lagal

Ram Publishing Company
P. O. Box 38464 • Dallas, Texas 75238

ISBN 0-915920-21-2
Library of Congress Catalog Card No. 75-44706

First printing 1976; Fourth printing October, 1979.

Printed by
Yaquinto Printing Co., Inc.
4809 S. Westmoreland
Dallas, Texas 75237

Printed in U.S.A.

I dedicate this book to my best friend, "Skipper," who has helped guide my ship of life through many stormy seas.

Roy Lagal

Other Publications:

Complete Book of Competition Treasure Hunting (The)
Complete VLF-TR Metal Detector Handbook (The)
Detector Owner's Field Manual
Electronic Prospecting
Gold Panning is Easy
"How to Test" Detector Field Guide
Journals of El Dorado (The)
Professional Treasure Hunter
Successful Coin Hunting
Treasure Hunter's Manual #6
Treasure Hunter's Manual #7
Treasure Hunting Pays Off!

CONTENTS

FOREWORD

Roy Lagal is a man who freely gives from his experience, his mind and his heart. Roy has lived a lifetime in God's great outdoors. There are few steeper mountains than those he has climbed and few hotter deserts than those he has explored. Always with him were Gerri and Larry, his wife and son, and his detecting and prospecting equipment. Thus, what he writes comes from the best of teachers — experience.

Roy has a keen, inquisitive and uncluttered mind. He has developed the ability to THINK. . . an ability I wish I had. His thought processes are always at work. What is not good or is impractical, he discards; the good and practical he keeps.

Roy has a heart bigger than the largest mountain he has climbed. He loves all creation, especially his fellow man. I have seen him weep when his auto wheel accidentally struck a cat. When the cat got up and ran off, Roy sighed a breath of relief. Would God want man any other way?

Roy Lagal? I'm glad I know him. And, I'm glad to know this much needed book, the *DETECTOR OWNER'S FIELD MANUAL*, is Roy's book. It comes from his experience, his mind and his heart — to you.

Charles Garrett

The author with his family in Lewiston, Idaho. Roy's experience with detector instrumentation began when he operated mine detectors during World War II. After his discharge from the Army in 1944 he married Gerri, and they have been on the treasure trail ever since. Their son, Lawrence, cut his teeth on metal detectors and was brought up prospecting and treasure hunting in our great outdoors.

PREFACE

Many books pertaining to treasure hunting have been published. They refer to TH-ing as the world's most fascinating and interesting outdoor hobby. Right they are! Many of these books are large treasure hunting manuals loaded with a wealth of information. Some are handbooks written by experts explaining various detector circuits and designs. These are invaluable to both the beginner and professional. A few "treasure" books very obviously exploit some particular manufacturer's brand name. This type of "treasure guide" can be spotted quickly as no other brand name or type of detector is shown or explained. POSITIVELY NO ONE PARTICULAR TYPE of detector will perform ALL the various jobs of treasure hunting and prospecting.

Each type has its good and bad points under certain field conditions and intended job use. There has long been a need for a field manual that explains not only the generalities of treasure hunting but also the many and varied aspects of successful prospecting and recreational mining using the electronic metal/mineral detector.

This *DETECTOR OWNER'S FIELD MANUAL* is the first book of its kind to give STEP-BY-STEP instructions in the use of metal/mineral detectors for treasure hunting, prospecting, recreational mining, PLUS complete instructions for the testing and evaluation of detectors before buying. The various types of detectors are correlated to actual field conditions so that you may effectively weed out unproductive detectors and ineffective methods. Clearly defined steps are set forth so that you may conduct your own detector tests, select the best type or types for yourself, and become more successful when using them.

I have been totally impartial in my discussions of the different detector types and make no mention of manufacturer, brand or model. Recommendations regarding choices for best field performances are based on information backed by solid electronic facts and proved success by professionals. Model or brand name is not all-important, BUT careful attention to selecting the TYPE of detector for your particular job definitely is. This often-overlooked fact alone has caused many treasure hunters and prospectors to quit in disgust before achieving success. Some types of detectors perform some jobs better than other types. Also, there are types that perform better than others in mineralized soil conditions. Careful attention must be given to all factors when selecting the TYPE of detector you will use and depend upon.

I heartily concur with the idea of trying before buying. This is not always possible, but let me recommend your local dealer, nevertheless. Most detector dealers are honest and competent. Some are more experienced than others. If you feel that your local dealer is biased as to brand or manufacturer or perhaps just inexperienced, I suggest you invite him to read this *DETECTOR OWNER'S FIELD MANUAL.* Ask him to help you perform some of the tests described in this book. Either you will find him to be prejudiced toward the particular line he sells and interested only in lining his own pockets, or you will find him receptive to letting you conduct your own tests in comparing detector types. In either case, the responsible manufacturer and dedicated dealer will come out on top and you will have the advantage of selecting the correct type of instrument for your job. Quality and performance should always be your constant goal.

My intent has been to present the complete facts to you — the consumer, the hobbyist, the treasure hunter, the prospector, the coin hunter, the relic collector, all those interested in the recovery of the lost — so that you will have the knowledge you need to pursue the many facets of treasure hunting and to select the equipment you require. It is my sincere hope that I have succeeded.

This *DETECTOR OWNER'S FIELD MANUAL* deals primarily with BFO's and TR's, as well as with the original VLF ground canceling types of detectors, those manufactured prior to the development of the VFL/TR discriminating circuit detectors.

As a final word, I offer my appreciation to the staff members of Ram Publishing Company to whom I am deeply indebted for their valuable assistance in preparation of this book for publication. My special thanks to Bettye Nelson, Editor. Without her help the book would not have been possible. Thank you very much.

Roy Lagal

BY THE SAME AUTHOR
AND THE SAME PUBLISHER

DETECTOR OWNER'S FIELD MANUAL
GOLD PANNING IS EASY
HOW TO TEST "Before Buying" DETECTOR FIELD GUIDE (VLF/TR-BFO-TR)
HOW TO USE "After Buying" DETECTOR FIELD GUIDE (BFO-TR)
THE COMPLETE VLF-TR METAL DETECTOR HANDBOOK (All About Ground Canceling Metal Detectors) (with Charles Garrett)
ELECTRONIC PROSPECTING (with Charles Garrett and Bob Grant)

NOTES ABOUT THE AUTHOR

Raised in Oklahoma on stories told by a great-uncle, an old Colorado and California gold miner, Roy Lagal has been a confirmed treasure/relic hunter since he was a small boy. When he was only eight, his great-uncle died and Roy was given his two "Doodle Bugs" (old-time dowsing and treasure-seeking devices). In his early years he had the opportunity to hear many buried loot stories and had firsthand experience with "Doodle Buggers" searching for buried money around his grandfather's Kansas farm. Though the early years spent following his "Doodle Bugs" over the Osage and Cherokee Nation country where he grew up provided neither junk nor money, Roy's interest in searching for the lost was heightened when during World War II he was assigned to work with mine detectors and other electronic devices.

After the War he accumulated a stake and in 1945 set off with his wife, Gerri, to find the Lost Dutchman Mine in Arizona. This venture did not prove successful, but later, with another stake at hand, Gerri and Roy traveled, prospected, dredged for gold, and treasure hunted over many states — the Texas/New Mexico desert areas, most of the West, and even Florida. They explored and lived in many

They finally settled in Lewiston, Idaho, where Roy started placer mining and continued to cache hunt. In Roy's own words, "I have never regretted it. All this Northwest is rich in Indian lore, gold rushes, and early day homesteading." By 1964 Roy had decided that yesterday's junk had become today's treasure — indeed, the New Gold — and that he would devote full time to digging bottles, relics, coins, and caches and combine it all into the antique business. As Roy says, "A little junk hunting never hurt anybody. The secret of professional treasure hunting is the ability to locate and recover yesterday's discarded or lost items, marketing them to collectors, museums and antique dealers."

Today, a real professional in all phases of treasure hunting (he won the 1966 National Treasure Hunter Champion title at the Oklahoma National Treasure Hunting Championship), Roy is one of the few who successfully specialize in the recovery of Indian caches, a very difficult field. Extensive knowledge of Indian lore and customs is required, as well as the ability to think as the older Indians did and the realization that many Indian caches were "buried for keeps" with no intention that the white man's evil gold should ever be found. Roy states treasure hunting has been good to him, and he would just like ". . . to stick around long enough to hit that big one."

(For additional information on the author's background, long-standing interest and experience in treasure hunting and details of Indian cache hunting, refer to the article, "Junk Hunting for a Living" by Roy Lagal, in the 1971-72 Edition of A.T. Evans' *TREASURE HUNTER'S YEARBOOK.)*

SECTION I

Detector Types and Useful Applications

Many different models are manufactured in each detector type. They vary greatly in quality, performance and price. Various detector types are: Transmitter-Receiver (TR); Induction Balance (IB); Radio Frequency (RF); Beat Frequency Oscillator (BFO); Phase Readout Gradiometer (PRG); Pulse Induction (PI). The Very Low Frequency (VLF) type includes such trade designations as Ground Exclusion Balance (GEB), Total Ground Cancellation (TGC), Magnetic Phase Deepseeker (MPD), Magnum Deepseeker (MD), Mineral Free (MF), plus many more such titles to denote their Very Low Frequency method of operation. There are the popular discriminators with the ability to REJECT most iron and other junk targets when searching for coins. This discriminating feature can be built into most types of metal/mineral detectors.

There may be additions to this listing, but basically they will fall under these TYPES and are merely phrases with which a manufacturer calls attention to some particular feature or operating characteristic. A complete list of types and a description of their operating capabilities follows, in alphabetical order.

CHAPTER 1

The BFO (Beat Frequency Oscillator) Detector

The BFO type is a TRUE metal/mineral detector. It is most commonly referred to as the only ALL-PURPOSE detector type currently manufactured. Of course, this does not mean that the BFO will perform ALL of the tasks assigned to it as well as might another type. It means that the BFO excells in some fields of operation and will perform with reasonable efficiency in all of the others. The newest VLF types will also do this. The versatility which makes it possible for the all-purpose BFO detector to operate under all field conditions will keep it a top favorite of treasure hunters and prospectors.

A BFO may also incorporate a discriminating circuit as an aid to rejecting iron, bottlecaps and other unwanted items when searching for valuable coins. The discriminating mode of operation has NO value, however, when used in metal *vs.* mineral identification. In fact, the meter indication will be opposite and will erroneously indicate a conductive ore specimen to be "bad." The same result will occur on a small natural gold nugget. For prospecting or metal *vs.* mineral identification the TRUE and normal mode of the BFO *can be used.* This is one advantage of the BFO type detector. It will always respond correctly and identify an ore sample as having a predominant amount of mineral (Fe_3O_4) or metal (any conductive substance). The BFO also has a fixed center of tuning and WIDE dynamic operating range. The electromagnetic field pattern produced by the BFO searchcoil is absolutely uniform; the TR (IB) searchcoil, *wound in any type of configuration*, will produce little or NO response on small gold flakes and may give an erroneous reading on marginal (though rich) ore specimens. The TR's inability to identify metal *vs.* mineral correctly can be attributed to

ORE SAMPLE TESTING with the BFO metal detector. Because of uniformity of detection over the entire searchcoil surface the BFO gives 100%-reliable metal/mineral identification.

the searchcoil's having two or more windings, the transmitter and the receiver sections of the loop, plus a very narrow (quick) dynamic operating range. In the ore identification field the BFO and certain VLF types are the MASTERS of all different metal detector types.

The BFO type does NOT penetrate black sand (Fe_3O_4) as readily as do the VLF and Pulse Induction types, BUT it can be called in to back up those two types where mineral (Fe_3O_4) exists (over approximately 90% of the earth's surface) IF THERE IS ANY DOUBT as to the identification of the target as metal or mineral. This includes searching for large ore veins, stringers, and pockets inside mines or caves; highgrading ore dumps; nugget hunting in mineralized streams; metal *vs.* mineral identification; *etc.* The BFO type does not produce on coins quite as effectively as do TR models, but it does operate reasonably well under ALL conditions with ease and stability. Many professional coin hunters (read the current magazines) ACTUALLY PREFER a BFO type for coin hunting due to its ease of operation over uneven ground. True, the BFO does take slightly longer to master, but many operators can identify most of their targets as to kind or depth before digging. There is no perfect detector, of course, and the operator's knowledge of different types, his experience and patience can mean the difference between success and failure.

The BFO types will not detect as deeply, especially in mineralized ground, as will VLF or Pulse Induction types, BUT the BFO type does NOT respond to tiny pieces of man-made ferrous iron. However, the VLF and Pulse Induction are considered to be "all metal" detector types and are SUPER-SENSITIVE to ferrous iron, regardless of the size of the pieces. The BFO does not have the sharp audible "whap" response of the NARROW DYNAMIC OPERATING RANGE TR types. Make no mistake, there are many facets to the treasure hunting and prospecting hobbies, and somewhere in all the re-

SEARCHING FOR MISSED MINE POCKETS AND VEINS. Pockets and small ore bodies in some old mines have occasionally been missed only by inches. In the next decade many millions of dollars worth of precious metals will be found by electronic detection methods.

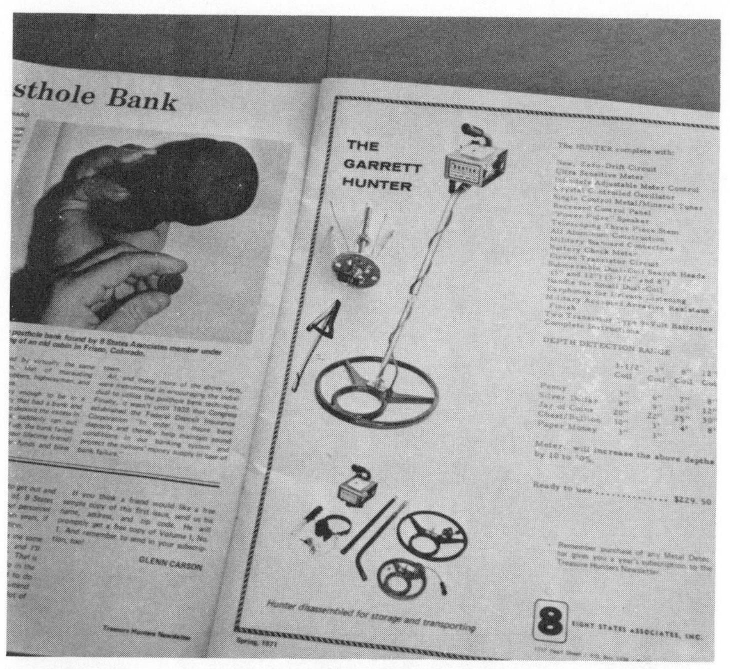

My thanks to Karl von Mueller of Segundo, Colorado, for the use of this photograph. With the aid of a BFO detector a treasure hunter found this diamond stickpin. The value of the diamond has not been revealed.

quired methods of searching there is sure to be one job that a certain detector type will perform better than another will. Do not condemn the detector type, however. Just correctly select the type you need and you will find yourself closer to success.

Treasure hunting is generally considered to consist of searching for large chests and money or precious metal caches. This is not necessarily so; many caches are small, perhaps only a small can or jar that may contain many thousands of dollars. Consequently, you need to employ a detector that is able to make use of the larger searchcoils. They are needed especially where small, unwanted metallic trash exists. Use of the all-purpose BFO will permit you to change to larger searchcoils in order to by-pass some of the small, troublesome signals. This is helpful when you believe that the cache should be a large kettle.

Veteran treasure hunters Charles Garrett and Ray Smith search this old home site for relics and treasures of the past. Charles is pointing to a large rock in the chimney that could easily conceal a cache.

There would be no advantage to digging all the small, worthless targets. If a cache is reputed to be, say, the size of a small tobacco can, the larger BFO coils will always be the best size with which to search.

Some BFO types make use of independently operated dual coils, a definite advantage since you may switch from a larger to a smaller coil to determine if the target lies near the surface and approximately how large and how deep it may be. This is a tremendous help to the treasure hunter who is searching for a metallic target larger than a coin or small piece of metal.

As noted, a few manufacturers have incorporated a discriminating circuit in their BFO types. This is helpful as you can, by flipping a switch, make your all-purpose BFO double as a "DISCRIMINATOR" that will reject worthless trash. Then, by flipping the switch back to NORMAL operation you regain full metal *vs.* mineral identification capabilities. In either flip switch position you can use different size searchcoils, from the small three-inch to the large twenty-four-inch deepseekers.

In all treasure hunts that involve the recovery of caches, remember that the kinds of locations and search areas will vary greatly. Perhaps one location will be around an old homestead littered with small nails and ferrous trash (the VLF and Pulse Induction types are super-sensitive to these items); another, in a wet, mineralized marshy area (here the standard TR type becomes more or less useless as the electromagnetic field pattern is disturbed by the wet mineralized soil); yet another, in a completely mineral-free area (here all types perform well). In specific job situations and under some soil conditions, the standard TR, VLF, Pulse Induction and PRG types will outperform the BFO types. However, remember the BFO and VLF are the *only* types that will perform many jobs perfectly AND still do all the others with some reasonable degree of efficiency.

If you have carefully studied the types of detectors and understand how each performs specific tasks better than others, you have also probably realized that the SPECIALLY DESIGNED AND ENGINEERED TYPES ARE GENERALLY LIMITED TO HIGH PERFORMANCE IN ONE FIELD OF ENDEAVOR ONLY. Perhaps this is why all professional treasure hunters and prospectors tend to select the more versatile instruments unless they have certain need for the specially designed types.

CHAPTER 2

The PI (Pulse Induction) Detector

The PI is another specially designed detector type. The PI has approximately the same sensitivity as all the VLF types. It operates perfectly over highly mineralized soil, but will produce the same false signals on hot (highly mineralized) rocks that are out-of-place as will all the VLF types. It is light and will operate perfectly in salt water. It produces an audible signal much the same as a high-pitched bell that is constantly ringing. It is more difficult to pinpoint with than the VLF types due to its continuous "bell ringing" response. It does not respond to small, natural gold nuggets or the lesser amounts of conductivity in most ore bodies. It cannot correctly identify metal *vs.* mineral.

The pulse induction type instrument, imported from England, is of excellent quality, lightweight, stable, and easy to operate. It is used in many different countries for security purposes and detection of weapons. The cost is not prohibitive, but there are some limitations to its usage. It is super-sensitive to small pieces of ferrous iron (nails, *etc.*). This makes it impossible to operate around old buildings where small trash abounds, but the VLF type has the same problem.

The pulse induction type detector should be used in highly mineralized zones and along salt-saturated beaches. It excells under these conditions, provided the beach is NOT littered with small ferrous trash. It may be used in treasure searching if the area of search is reasonably free from ferrous trash. It has the advantage of being free from the troublesome tinfoil indications common to most detectors. It is the ONLY detector capable of locating a man-made ferrous target (tin can or metal box) in an ore dump that contains BOTH negative (magnetic iron) and positive (conductive) low grade ore; the instrument simply does not see either. Thus, the pulse induction type is probably best used by highly skilled operators for successful coin hunting in heavily mineralized soil and along ocean beaches *where trash is not abundant.*

CHAPTER 3

The PRG (Phase Readout Gradiometer) Detector

Without any question, this type is the king among discriminating detectors. It employs a gradiometer sensor and operates about the same as does a TR type. It has about equal sensitivity to the VLF and Pulse Induction types. Salt water does not affect it and it is unbeatable on a beach that does not contain magnetic black sand. There is, however, one major drawback to the PRG. If the soil contains moderate or high amounts of Fe_3O_4, magnetic iron — approximately 90% of U.S. soil does — the PRG is almost *impossible to operate.* Another drawback may be its cost. Cost should NOT be considered, however, when choosing a quality instrument. Don't misunderstand; the highest priced detector MAY DEFINITELY NOT BE THE BEST DETECTOR FOR YOU! The PRG is rather heavy (about seven pounds), but it is stable, sensitive and gives almost perfect readouts on targets in non-mineralized search areas. If you are coin or treasure hunting in an area free of iron mineralization and can afford the price, you should test this instrument. The PRG is designed to operate in salt water zones and over non-mineralized ground with excellent target identification.

Consider and evaluate all evidence before purchase. If you truly have use for these specially constructed instruments, they are fantastic; if not, select your detector from types that have more versatility. Additional technical information may be obtained from reading *THE ELECTRONIC METAL DETECTOR HANDBOOK* by E. S. "Rocky" LeGaye and *SUCCESSFUL COIN HUNTING* by Charles Garrett. (These books are listed in the back of this book.)

CHAPTER 4

The RF (Radio Frequency) Two-Box Detector

The large two-box detector is actually a transmitter-receiver type, commonly referred to as a deepseeker. (Until recent development of the VLF type in the horizontal loop style, the two-box RF was rated as the deepest-seeking detector on the market.) As with any deep-seeking detector, the two-box type has advantages and disadvantages. It should be carried as closely to the ground as possible to assure the utmost in added depth penetration and permit the detection of smaller targets. A carrying strap or extension handle is always provided for this purpose. The two-box will generally not detect a metallic target smaller than the size of a baseball. An experienced operator (using a two-box detector in the absence of mineralization) might find it possible to detect an object as small as a silver dollar.

This type detector is used to locate buried pipes, cables, magnetic or non-magnetic ore bodies and other large metal objects. Some special two-box RF models are manufactured specifically to locate pipes and cables. Commonly called "pipe finders," they incorporate the use of a wire that is connected between the transmitter box and the pipe or cable. The receiver box is then carried along the ground above the buried portion of the pipe or cable. The detector picks up the signal being transmitted through pipe or cable. For this purpose the two-box RF has NO EQUAL. These commercial pipe locators are usually sold by a cable or pipe supply company, not by detector dealers. Some metal detector dealers still stock the regular two-box deep-seeker RF without the pipe and cable hookup. This model has been used for years in attempts to locate magnetic or non-magnetic ore bodies.

The complete inability of any type transmitter-

One of the oldest type commercial detectors ever used . . . the two-box RF transmitter-receiver. This type detector when equipped with a small cable hookup for transmission of electrical energy through a conductive iron pipe is known as a pipe or cable locator. Though it is a deep-seeking detector, when used for treasure hunting it has many limitations. Ray Smith is searching for a post hole bank along this fence line. He must hold the instrument above the grass to prevent static from the unshielded coils contained in each box. A VLF type equipped with the larger searchcoils would have performed better under these circumstances. It is sometimes difficult to determine the exact size of a cache until it is recovered and the RF two-box will not respond to smaller targets. Thus, there may be some doubt whether you have missed the cache for which you were looking.

receiver to *identify* metallic ore correctly in the presence of high mineralization (Fe_3O_4) has resulted in many failures and disappointments for miners. This type metal detector is not a successful aid for prospectors. Many failures have been directly traced to the use of the two-box RF detector. Actually, however, these failures should be attributed to poor judgment in selecting the TYPE of detector required for a certain job. When the two-box detector is used for ore body detection, it must be remembered that it may respond either to metal or mineral. Any damp rock containing a high degree of mineralization (Fe_3O_4) may respond the same as metal. You will find supporting evidence of this characteristic from the many empty holes dug because of false readings received in areas of high mineralization. It is true that this type of detector has been successful on a few occasions, but it may never be ascertained whether success can be attributed to the operator's correctly identifying the target or simply to haphazard digging wherever the instrument responded!

Despite the problems, the two-box RF has been used for years by many treasure hunters. Of course, it must be remembered that if you previously wished a deepseeker you were practically forced to use this unit. Many deep and empty holes have been dug and abandoned all over the world. Almost any underground wet spot can cause a reasonable target indication even while the detector is tuned in the metal mode of operation. HOT rocks (containing Fe_3O_4) mixed in among other, less mineralized rocks will almost always result in signals which cause useless digging. In many states with soil of a high mineral content treasure hunters have been forced to abandon searches due to numerous false signals and poor depth penetration in the mineralized soil.

The new VLF detectors with the horizontal loop configuration still produce the same false responses on hot or highly mineralized out-of-place rocks, but they have the advantage of responding to smaller targets. They also penetrate the mineralized soil perfectly, plus there is a choice of searchcoil sizes. Thus, there are many advantages over the two-box RF. The VLF deepseeker will double as a nugget shooter and coin hunter.

I cannot overstress that the beginner should ALWAYS carefully consider what his major interest will be and then choose the type of detector that performs best in that particular field. The two-box RF is excellent for pipe and cable location, also for location of larger targets in non-mineralized soil; but full investigation of the facts should be made before considering it for any other phase of treasure hunting. Careful testing of both types, the two-box RF and the VLF, will quickly convince you of the tremendous advantage offered by the horizontal loop style VLF. No one can quarrel with field test results, and the outcome will probably save you money . . . or perhaps make some for you.

CHAPTER 5

The TR or IB (Transmitter-Receiver, Induction Balance) Detector

For simplicity, we will hereafter refer to the transmitter-receiver, induction balance detector as "TR" since, for all practical purposes, they are the same. The horizontal loop TR is very good in the *coin hunting field.* However, due to erratic operation resulting when the searchcoil is moved up and down over highly mineralized soil, the TR leaves a lot to be desired for all-purpose treasure hunting and prospecting. The larger the searchcoil, the more erratically and "quickly" the TR responds to the mineralization content of the soil. This "quickness" of response is *excellent* and *desirable* in the search for coins, but presents a drawback when stable operation is necessary for searching over uneven ground, rocky areas, *etc.* In these situations the TR cannot be kept tuned at its highest peak of operational efficiency.

Many manufacturers have attempted to alleviate the problem by adding auto-tuning, push-button retuning and mineralized ground control features. Any type of ground control or ground adjustment feature on a standard TR is nothing more than a gain control or sensitivity adjustment. When you use the control to lessen the effect of mineralized ground, you also lessen the sensitivity. This, of course, is necessary in certain areas. It does not solve the problem, but decreased sensitivity does make detector operation easier. The auto-tuning feature places the detector in a completely automatic mode. The tuning never has to be adjusted; merely turn the instrument on and off. At first this seems fantastic, but it has a drawback: the audio tends to "overshoot" as the auto-mode keeps trying to adjust the tuning in accordance with the metal or mineral background. Because of this, the auto-mode should be thoroughly tested before using as the loss in stability and sensitivity may be too great. The push-button re-tuning system is similar, but it has no effect on stability or sensitivity. The detector is tuned in a normal manner while depressing the push-button control; the control is then released and the detector returns to its normal mode. When the tuning needs adjustment to compensate for ground or temperature changes, press the push-button control and the tuning is instantly electronically reset to your original point. This method has absolutely no effect on stability or sensitivity. To operate in the completely automatic mode simply hold

John Quade, movie actor, discusses metal detectors with Bert Green, "Carolina Ernie" Curlee, Stuart Auerbach, Claude Hooks, and George Mroczkowski. The fellows were getting ready to participate in field trials sponsored by Garrett Electronics. All types of detectors have been used in field trials. The choice depends a lot upon the type of targets buried and their depth. For an in-depth look at field trials and how to conduct one, write to Ernie for information about purchasing his new book (HOW TO SPONSOR AN ORGANIZED TREASURE HUNT, $5,Treasure House Publishing Company, 2918 Dunaire Drive, Charlotte, NC 28205). It is rewarding reading I recommend for every treasure hunter who has attended or who plans to attend even one of the many hunts that are held each year.

the push-button control in a depressed position. Now, however, the operation will become slightly erratic over mineralized ground and some sensitivity may be lost.

Many TR detectors are designed with a DISCRIMINATING mode in addition to the NORMAL mode of operation. This is sometimes referred to as a hybrid or twin circuit. It permits the detector to be used in the normal mode for relic, junk and cache hunting and, of course, for coin hunting. In the normal mode all metallic targets are acceptable. Usually a switch or adjustment control is provided to permit you to energize the discriminating circuit. With discrimination you may reject many junk or unwanted trash items, such as tinfoil, bottlecaps and small ferrous iron pieces. The discrimination control on many models may be adjusted so that the detector will reject pulltabs. The pulltab has a high conductivity factor and the increased amount of rejection would drastically cut sensitivity, plus causing you to "lose" nickels and small rings. Whether the loss in sensitivity necessary to accomplish pulltab discrimination is practical depends on the search area and the operator's desire. As previously discussed, the twin circuit can always be returned to the normal mode for all-metal detection.

Searchcoil construction also affects these dual-purpose instruments. In order to achieve acceptable discrimination the searchcoil configuration sometimes has to be changed slightly from the standard TR design. Some TR detectors that feature discrimination are more adversely affected by mineralized ground than are non-discriminating types. All these points should be taken into consideration and evaluated before purchase of a detector. The twin circuit or dual-purpose unit is more practical for the average hobbyist who owns only one detector, but the competitor who enters speed contests or treasure meets and has no need for discrimination might desire the normal TR detector with a 2D type searchcoil because of its easier operation over mineralized ground.

One-hundred-percent-accurate metal *vs.* mineral identification is NOT POSSIBLE with a TR due to the configuration of the searchcoil. When tuned in the metal mode, the transmitter portion of the searchcoil sees negative targets (mineral iron, Fe_3O_4) as positive; the receiver portion sees the mineral correctly as a negative target. This, however, presents a problem as there is no foolproof way to determine with which part of the searchcoil the target may come in contact. You can test this by placing a coin-sized amount of magnetic black sand and a coin several inches apart under a piece of newspaper. Then tune your TR and mop or slide the coil across the newspaper just as you would when searching outdoors. At some point(s) you will get approximately the same indication on both targets. Now you understand what happens when you attempt to use the TR for small nugget searching and metal *vs.* mineral identification. The TR circuits are simply overloaded and produce erroneous responses. A small highly mineralized pebble may overload the TR's circuit due to the configuration of the searchcoil and respond the same as would a small coin.

Test the TR's response on small amounts or thin flakes of natural gold. You will discover the TR generally does not respond to these marginal targets. If the TR searchcoil is not properly Faraday-shielded it may respond to the tips of your *fingers.* Always check this because a coil that is not shielded or that is incorrectly shielded will give erroneous signals when touched by weeds or grass.

Larger searchcoils are difficult to use on the TR because the solid-type construction presents a weight problem. Because of this factor most manufacturers do not promote their large TR coils, preferring to sell the TR instrument for the purpose for which it is designed . . . fast operation and good penetration on COINS.

It is an undisputed fact that many caches have been found with the TR over the years. Also, many LARGE natural gold nuggets MAY have been recovered, using the standard high frequency types of TR's. Yet the fact remains that because of the TR's quick response operating characteristic the TR searchcoil does not respond correctly and is erratic when used in highly mineralized zones. The TR is completely inhibited inside mines or caves that contain mineralization combined with damp matrix. The wet, mineralized soil and rock disturb the electromagnetic field produced by the searchcoil and limit the coil's operating capabilities. Wet marshy areas will do the same thing. Try placing a coin only one inch deep in a wet, rainsoaked, mineralized area and notice the *reduction in the detection signal.*

The TR excells in ghost-towning and battlefield searching in non-mineralized or neutral ground, for here variations in searchcoil operating height are of little consequence. It will out-perform almost all other types of detectors in coin hunting, bowing only to the VLF and pulse induction types in highly mineralized ground.

CHAPTER 6

The VLF (Very Low Frequency) TR Detector

This is the latest mineral-free operation, magnetic phase detector in the TR line. A variety of trade names call attention to its super-sensitivity and mineral-free operation — GEB, TGC, MF, MPD, and many more such designations. The VLF type has a ground or terrain control to zero-out the effects of mineralized soil or rocks. Of course, it will still respond "positive" on out-of-place hot rocks (any rock containing *MORE* mineral content than the matrix to which the detector is tuned). This response is troublesome in some areas, but the unit will not falsely indicate unless the hot rock is rather close to the searchcoil. The VLF type has rapidly become famous for its deep-seeking capabilities. With the use of average size searchcoils, the VLF types will detect coins and small gold nuggets at unbelievable depths *even through the most highly mineralized rock or soil.* Fantastic? Yes! Yet it, too, has faults. It is super-sensitive to small pieces of ferrous iron. If the search area is littered with small trash it is practically impossible to operate effectively. The availability of large searchcoils for the VLF have made the two-box RF model detector obsolete for deep cache searching. The VLF type will go just as deeply (using LARGE coils) and has the advantage of responding to smaller targets! It can also utilize different methods of discrimination, and it is NOT affected by mineral content (magnetic iron), as is the two-box RF deepseeker. These features give it many great advantages over other types of detectors.

The VLF type is as sensitive, or more so, than the PRG and pulse induction types. In addition, it has the advantage of operating almost perfectly over mineralized ground where the PRG is nearly useless. The VLF has

The author and Tommy T. Long. Mr. Long, of Boise, Idaho, has been a hunting companion of the author for many years and is now a business partner. Tommy T. hunted this same 120-year-old park with the author nineteen years ago. The new VLF type detectors being used, while causing the recovery of much worthless ferrous junk, were responsible also for the recovery of many small, valuable Barber dimes too deep for previous detection with other types of detectors. Actually, patience, practice and familiarity with these VLF types will soon eliminate much of the useless digging. The use of VLF types is almost mandatory for the recovery of deep coins, especially in mineralized soil. Many older parks are now being reworked by VLF operators.

the capability of pinpointing the target MUCH more accurately than the pulse induction type because the "after-ring" of the PI type confuses the operator. The VLF type can be used to search for conductive ore veins and will operate perfectly inside mines and caves.

You must remember that when it is used to search for valuable ores it will also respond to non-conductive or magnetic bodies that are out-of-place. (This is true even though the detector has discriminating capabilities that use either the low frequency TR or true mineral-free method of discrimination.) For this reason you MUST use a BFO type or use one of the newest full capability VLF's that can correctly identify metal *vs.* mineral. The BFO does not penetrate magnetic rock as well as the VLF, but the hot rock that caused the VLF to respond erroneously is generally never very deep. If the BFO calls the target "mineral," you know it was only a false indication. If the BFO does not respond *either way,* the target will probably turn out to be a conductive (rich) ore deposit beyond the depth limitations of the BFO. If the conductive pocket or vein is near the surface the BFO will indicate "metallic." For the first time in history it is absolutely possible to conduct mine surveys with electronic detectors and be able to penetrate the mineral matrix. The electronic locating of ore

Charlie Weaver, Lewiston, Idaho, uses a VLF type detector to rework a very old city park that has been "cleaned" countless times. Notice the depth necessary for recovery of these deeper coins. Plugging the turf is not the preferred or best type of recovery technique, and it should be done only when the soil is wet enough to allow the grass to resume growth. Thus, the plugging method should almost always be used only under winter search conditions. Take great care to replace the sod and firm it down so the grass will reroot.

A vacant lot can produce many surprises! Fred Mott started out to find an old bottle dump . . . and ended up finding some rare relics, many of them buried quite deeply. Fred is using a VLF type instrument. The irregular ground surface did not affect the VLF due to its mineral-free operation. It is also super-sensitive to both small or large targets, a fact which prevents overlooking old coins when relic hunting. Bottle dump searching is best accomplished by the use of the larger searchcoils available for these super-sensitive detectors. The larger coil helps by-pass some of the smaller targets when dump hunting, saving considerable time. The same would hold true when cache hunting.

stringers and missed pockets of high grade ore will eventually make the VLF type the most productive of any detector used in the mining industry. Remember, however, if your VLF detector is not a "full capability" instrument the BFO *must* be used to identify the deposit correctly. Failure to utilize a BFO for this purpose will result in a great amount of useless expenditure of labor and money for exploration work.

Because of the very low frequency operation and type of searchcoil construction, some VLF types are affected by power lines, electric lights and other electrical interference. Erratic meter and audio response will result. Careful inspection before buying and comparison of the VLF types will protect you against this unnecessary fault. There are many different types of searchcoils. That is to say, they are wound in many different configurations, both co-planar and co-axial. Some will pinpoint better than others; some cause erratic operation; some prevent interference from outside electrical power sources. All these factors must be taken into consideration when contemplating the purchase of one of these advanced-design instruments.

The VLF type should NOT be confused with the standard high frequency TR (IB) models.

Here is big Tony rapidly whipping the Red Baron (manufacturer's designation, S.P.D.) searchcoil over a target he has just detected. Failure to hold the handle button in the depressed position or to maintain a certain whipping speed range, neither too slow nor too fast, may result in loss of signal or incorrect identification of targets. Reasonable identification can be accomplished on all targets, however, except bottlecaps. Bottlecaps are almost impossible to discriminate against using this type VLF ground canceling method. This partial method of discrimination in the mineral-free mode reminds one somewhat of operating a true TR discriminator in the completely automatic mode. This method is considered by some to be quite successful in certain areas; however, as you can see from Tony's efforts, it could become quite tiring.

Allan Cannon, Pomeroy, Washington, is shown working in one of the large Northwestern lumber mills. He is using a VLF type detector to locate nails or other metallic objects which may be imbedded in the logs. If these objects are not located before milling they could cause a great deal of costly damage to expensive saw blades. TR or BFO type detectors are not recommended for locating elongated ferrous objects which lie at approximately a 90°-angle to the searchcoils. The degree of angle and the mineralized sap in the green logs combine to make it virtually impossible to achieve detection of the metal objects with the TR or BFO. Since the VLF is not hampered by the sap and the angle of the object facing the searchcoil, it is perfect for deep detection. The utilization of VLF type instrumentation for the purpose here described will result in saving countless dollars in sawmill operations.

You will probably find the Standard TR is fairly easy to use in the search for single coins if the ground is not too highly mineralized. The VLF, however, will detect these same coins at great depths through even the most highly mineralized soil, but it would also detect very small pieces of ferrous iron trash. This is very frustrating and causes most coin hunters using a non-discriminating VLF type to give up in disgust. If you select an area of search where trashy targets are not abundant, then the VLF will truly reach down to find the older, more valuable coins. Some older parks are very littered with trash; some are not. The ones that do not contain much trash really pay off for the operator of the non-discriminating types.

Any kind of discriminating ability is welcome on the VLF type detector when used for coin hunting. Of course, a true, mineral-free discriminating VLF detector that employs normal operating methods could solve the problem of unwanted trash while retaining full depth capabilities. However, due to design complexities, this type VLF may be long in arriving. The problem is greatly alleviated by some manufacturers who simply supply a method to switch from the mineral-free mode to a standard TR discriminating mode. This method allows full discrimination capabilities in mineral-free ground, but presents more problems than occur with the standard TR operated over high mineralization. This is because the very low frequency of the VLF types is more adversely affected by mineralization than are standard high frequency TR's. Also, the searchcoils may be heavier and the battery drain may be greater than on standard coin hunting TR's. When selecting a VLF type for coin hunting in trashy areas, it would be wise to select one that has a proved coin hunting record.

Specialized instruments require more effort for the beginner to master. I suggest that one of the less complicated instruments be used as a back-up when problem situations arise. Probably the best choice to use in conjunction with the VLF type is the reliable all-purpose BFO. To do this will give the operator the advantage of having TWO completely different types of instruments at his command. The VLF is actually a TR type, and it would be a waste of money and time to use a standard TR as a companion instrument. When the purchase of an advanced design detector is considered, attempt to find a qualified dealer who can explain all the various types. Insist on seeing how they work. The salesman or dealer who discredits his competitors, insisting that there is *one* best type and that the type he stocks can do it ALL better than any other type, is ill-informed and unreliable. There can be a "best" where quality of construction and field performance are concerned, but NO specific *type* of metal detecting instrument can perform ALL the various jobs in treasure hunting and prospecting better than all the other types. This is why most families and ALL professional treasure seekers and prospectors own two or more different types. Fortunately, the increased use of metal detectors as a world-wide, profitable hobby and the formation of numerous treasure clubs has caused the public to become much better informed. A desire for knowledge and an open mind can spell the difference between success or failure.

CHAPTER 7

The Discriminators (Selectmatics, Rejectors, Analyzers, etc.)

These detectors are equipped with a conductivity "readout" indicator, either meter or audio. This means you may set a control (some are pre-set at the factory) to discriminate or reject metallic targets of a low conductivity factor but to accept as "good," targets of a higher conductivity. Ferrous targets (iron) generally have the lowest conductivity. This includes nails, bottlecaps, tinfoil and other unwanted junk frequently encountered while coin hunting. Non-ferrous targets having a higher conductivity factor, such as coins, aluminum, brass, *etc.*, will be accepted as "good."

A small problem arises when you adjust your detector to reject aluminum pulltabs. The detector will also reject nickels (which have poor conductivity) and some small rings. These rings are generally fourteen to twenty carat gold or silver. They have a small metallic mass effect on the detector's electromagnetic field. This fact, coupled with their small amount of impure alloys, will cause rejection. If you test a .999-fine silver bar you will notice that the discriminating circuit will accept it as good. Small rings made from sheet silver will be accepted while small rings of approximately fourteen to twenty carat gold or silver may be rejected. The handmade silver ring was more pure and had higher conductivity. Self-conducted tests can be very helpful in understanding discriminating detectors. The discriminator will indicate a small gold nugget as "bad," but will indicate small gold coins as "good." The rejection is caused by the smaller mass effect and the rounded edges of the gold nugget. Discriminators should NEVER be used in metal *vs.* mineral ore sample identification.

Most discriminating detectors have a control to per-

Willis "Pinky" Noble, Lewiston, Idaho, is reworking one of the older parks in that area. This particular park has been worked by thousands of experienced coin hunters but still continues to produce a few goodies that remain at great depths. Notice that "Pinky" is using an extra-large TR searchcoil. He states that he searches in the normal mode and then switches to the discriminating mode to identify the target correctly. This method prevents missing some of the deeper targets since ALL discriminating circuits that identify by the amount of conductivity of a metallic object must lose some of their sensitivity. Our thanks to "Pinky" for the photo.

mit adjustment of the amount of rejection desired. This arrangement is more desirable than that of the types that have been factory set. Factory-set discriminators *cannot* be adjusted to permit rejection of tin cans and other large metallic targets when you might wish to detect a known object such as a large silver bar, large aluminum mass, brass foot valve, *etc.*

Remember that ALL discriminating circuits MUST indicate as *bad* any soil that contains mineral (natural magnetic iron, Fe_3O_4). This presents an operational problem because the height of operation of the searchcoil over mineralized ground is very critical. Discriminating detectors without mineral-free capabilities have to be operated

Bob Boykin, Dallas, Texas, who heads one of the country's largest security forces, is searching for a bullet fired during a robbery. The bullet is needed for evidence. Bob is using a BFO type detector with small searchcoil. He says he has good success with this type detector as it is adaptable to and successful in the many difficult situations he encounters. Regardless of weeds, grass or uneven terrain, the BFO type is stable and easy to operate. Thank you, Mr. Boykin, for furnishing this photo.

at a constant height above mineralized soil; otherwise, the indicating meter will fluctuate so greatly that you cannot obtain a true reading. This fact also holds true if the instrument is one that discriminates or rejects by audio means. The audio type would simply by-pass some of the targets.

Detectors called "discriminators" are generally equipped with only ONE searchcoil and do not permit use of larger coils, a drawback when attempts are made to use these instruments for cache and relic searching. Also, they cannot be used for metal *vs.* mineral ore sample identification where either small or large coils may be required. The factory-set discriminator is generally adjusted to discriminate on ONLY ONE SEARCHCOIL.

Many manufacturers have overcome this disadvantage by employing a hybrid or twin circuit, allowing the detector to operate in the normal mode. A switch or adjustment control is provided whereby the instrument can be changed to operate in a discriminating mode, a very advantageous arrangement since it provides TWO instruments in one. There would be no need to purchase both a discriminator and a normally operating detector to make use of various sizes of searchcoils and various detector circuit capabilities. Some manufacturers also guarantee FULL AND ADJUSTABLE DISCRIMINATION ON ALL SIZE SEARCHCOILS WITH NO LOSS IN SENSITIVITY on some of their types and models. HOW THIS IS DONE is of little consequence to the buyer who needs to know only that IT IS POSSIBLE. Advantage should be taken of the new methods. Almost any TYPE of detector may be designed with a discriminating circuit, and the ones offering both NORMAL and DISCRIMINATING modes of operation are the best buy. Always keep in mind, however, that discriminators are not 100%-accurate. Many different factors enter into their operation . . . the mineral content of the soil, the conductivity factor of the target, and the expertise and experience of the operator.

Many buyers purchase a discriminator, expecting to dig nothing but coins. This is far from what may actually happen. First, you must purchase a quality instrument that is STABLE. This means, for one thing, that the instrument will actually give a very slight but discernable indication (either meter or audio) on deep targets. Detectors that drift will actually mask or "blank out" deep target signals and you will miss some good things. Second, you must practice in order to understand what the discriminator is telling you. The beginner may expect the meter to say simply "good" or "bad," and that the audio will be more pronounced over "good" targets. However, the meter indication is always very slight on deep targets.

If the instrument is stable these slight meter indications can be interpreted as good or bad. Some manufacturers and dealers fail to alert the operator that DEEPER TARGETS DO PRODUCE FAINT SIGNALS. This is one of the major reasons why attention to versatility and design details of discriminating detectors are all important.

A detector classed as a common discriminator, equipped only with one size searchcoil, has little value other than to hunt coins. Detector types that combine both the normal and the discriminating modes of operation, plus a full assortment of searchcoils, offer the treasure hunter and prospector the widest range of operation. Any type of detector offering this dual method of operation (both normal and discriminating modes) should

As unlikely as it may appear, many caches have been recovered from beneath front door steps. Because of the presence of many nails and small trash items in areas like this it is sometimes preferable to use the BFO rather than the super-sensitive, very low frequency type of detector. The same principle applies to searching buildings in which numerous nails were used.

employ full Faraday-shielded searchcoils to prevent false signals when operated in wet grass and weeds. It should be able to accommodate a full complement of searchcoil sizes, from the small nugget hunting coils to the deep-seekers.

A BFO type and a TR type react differently when in the discriminating mode. The TR generally has an adjustment whereby the amount of discrimination may be increased or decreased. The typical method for setting this control is to set it to reject a bottlecap when the cap is about one inch away from the bottom of the searchcoil. If you move the bottlecap closer the detector will accept it as good. While discrimination may be adjusted any way you wish, if you set the control to reject the bottlecap at a distance much more closely than one inch you will drastically *decrease* the sensitivity. Regardless of the discriminating control adjustment you make, the TR will STILL indicate the bottlecap as being good if it is brought into close contact with the searchcoil. This is a failure the TR cannot overcome. The circuit is simply OVERLOADED and produces a positive signal. Also, the TR has two or more windings in the search loop. One transmits; the other receives. So when a bad target is passed over at a shallow depth or the searchcoil actually makes contact with the target, the TR will indicate it as good. Of course, the standard TR discriminator produces better results than the BFO discriminator on deeper, coin-sized targets because the TR's audio circuit is easier to interpret.

At roadside picnic grounds there are usually many pulltabs and bottlecaps lying on the surface or just barely buried. They will produce positive signals. To insure that the detected object is worthless and not to be dug, the TR discriminator must be raised above the ground approximately one inch and passed back over the spot. The detector will have to be retuned each time. This procedure is greatly aided by the push-button control for instant retuning. You simply reset the tuning, scan over the spot, then reset the tuning after you lower the coil back to the ground for deeper searching. Extreme patience and slow searching procedures are sometimes required for surface or near-surface targets, but if the ground is littered heavily with trash searching isn't usually worth the effort.

TR detectors incorporating a completely automatic mode of tuning (referred to as auto-tuning) are completely helpless when in the discriminating mode of operation. When the target is passed over, regardless of good or bad or of the target's depth (provided that it is within detectable range), the detector responds with a positive "zap" or signal. This also happens when you pass in close contact with a highly mineralized pebble or rock. This is no fault of the unit; the auto-tuning system is simply trying to adjust the unit too quickly. Because of the speed with which this function must be performed, the effectiveness of the auto-tuning mode for discrimination purposes is completely destroyed.

The low cost model TR discriminators may remain, for a while, as popular coin hunters for beginners, though they do have a few restrictions – namely, in trashy areas where the coins are very shallow or over uneven, bumpy, mineralized ground where the searchcoil cannot be kept in contact with the ground or operated at an even, controlled height.

The BFO discriminating detector may be designed and manufactured to discriminate either by audio or by meter indication. The metered discrimination principle is unquestionably the more practical of the two methods. Only this method will be discussed in this chapter. In many specific places and under certain conditions the BFO discriminator has more practical value for coin discrimination than does the TR type. Consider again the roadside picnic ground. Coins, rings, bottlecaps and other metallic targets are generally only under the very top of the ground or lie at a very shallow depth. When the TR discriminator comes into close contact with the target, it produces a positive response or "blip" due to the nearness of the target, REGARDLESS OF WHETHER THE TARGET IS A COIN OR JUST PLAIN TRASH. The BFO does NOT have this reaction caused by close contact to metallic targets. The meter will correctly identify all targets, whether the searchcoil is touching the target or at the maximum detectable distance. Of course, there is the problem of constant audible signals (the same as with the TR) produced by numerous trash targets, but this can be easily overcome. Simply turn the audio down and observe the meter for correct discrimination indication. This method, while considerably slower, is much more efficient, especially if there are a great number of trash targets. (The TR cannot operate with meter indication *only* because its coil configuration produces good indications on shallow trash.)

The small strip along sidewalks can also be searched by the meter method. Many traveling treasure hunters simply give up on roadside parks and stopping places because of the great amount of trash. However, many good finds continue to be made by those who understand the proper methods of searching. Because one instrument may perform a particular job in a specific area better than another type, you can easily understand why many treasure hunters carry different types of detectors. Valuable rings and jewelry are the most common finds because persons who lost them did not remember where the items were lost or have time to return to search. (They probably did not have a metal detector, either.)

The VLF type detectors with trade names such as GEB, TGC, MPD, Magnums, *etc.*, are, in effect, discriminators. These VLF types can be adjusted to discriminate against or to exclude mineralized ground (magnetic black sand), but this does not mean they discriminate between good or bad metallic targets at the same time. (Actually, when in the mineral-free mode, they are super-sensitive to ferrous iron targets at all times.) When discrimination by amount of *target conductivity* is desired, the VLF generally has another circuit installed to permit the operator, by flipping a switch AND REDUCING SENSITIVITY, to return the operation to that of a standard TR type discriminator. Of course, in that case the VLF types cannot be operated *mineral-free* and no longer penetrate the mineralization freely. The VLF discriminator circuit is still operating in the extreme low frequency range, and *frequency* plays an important part in TR operation over mineralized ground.

A standard TR discriminator operating in the medium-to-high-frequency range may be fairly easily operated over ground containing a moderate amount of mineralization. When the VLF type discriminator is operating at a very low frequency (not in the mineral-free mode) over ground containing negative mineralization, it becomes *super-sensitive* to iron mineral and practically impossible to use. When a discriminating circuit is ADDED to a mineral-free detector, many advantages but also many problems are presented. This type detector designed with both circuits must have a switch to change from mineral-free operation to discriminating operation. A large percentage of ground in the U.S. contains negative mineralization (magnetic iron, Fe_3O_4), and if you attempt to use this low frequency discriminating mode in such areas the instrument will be extremely erratic. The sensitivity (gain) can be turned down to help eliminate the problem, but the sensitivity may have to be decreased to the point where it is LESS than the usable gain of a standard high frequency TR.

The discriminating circuit added to the VLF type with mineral-free operation (for discrimination OR mineral-free operation) should never be confused with standard TR discrimination. Neither should it be confused with TRUE *mineral-free* discrimination. However, in mineral-free areas the low frequency discrimination circuit will operate perfectly and is a definite asset. Understanding the difference between mineral-free discrimination and low frequency TR discrimination and then having the manufacturer or dealer explain it thoroughly and honestly will help you. You will not be led to erroneous conclusions nor expect your particular detector type to do jobs under conditions for which it was not designed. Quality and performance are all-important in selecting any detector. Careful attention as to type and intended job use is absolutely mandatory.

If you desire to further your understanding of VLF type detectors, I suggest you study my new, all-encompassing book, *THE COMPLETE VLF/TR METAL DETECTOR HANDBOOK (All About Ground Canceling Metal Detectors),* published by Ram Publishing Company. These new VLF detectors have, almost overnight, captured perhaps 90% of the higher priced detector market. Why? Because certain of them are, without question, the best, most all around type instruments manufactured. In my new book I fully explain the VLF's and their capabilities, as well as their shortcomings. You should not be without this information.

SECTION II

Treasure Hunting and the Metal/Mineral Detector

CHAPTER 8

Ghost-Towning

One of the most popular and rewarding hobbies in America is "Ghost-towning." This word refers to a number of activities. Discovering old coins, perhaps a buried treasure cache, relics or antiques dating back to the Pilgrims, and many other lost items of yesterday can be considered ghost-towning. Some items will have no value except for historical interest, but that alone is enough to rouse the curiosity of most treasure hunters.

Any place people have gathered, camped or lived will produce relics or coins. There are thousands of abandoned town sites, old forts, homesteads, farm house locations . . . the list is endless. If people were there in the past, chances are some old coins or nice relics can be found. Finding a place to search will never be your problem, just the time needed to pursue and enjoy your hobby. Remember that most of the surface items were picked up by relic and antique hunters long ago. You will need a good general-purpose metal detector because most objects will lie below ground surface or at least be concealed.

Ghost-towning differs from the prospecting hobby in that you may use almost any type of detector with reasonable success. The TR, VLF, BFO and pulse induction types will all, to some degree, find relics and coins. The PRG would operate perfectly in non-mineralized areas and have good identification readout on targets. The pulse induction and VLF types will not be affected by the minerals, but both present problems when used around old homesteads where trash abounds. Due to their supersensitivity to small ferrous (iron) targets they may sometimes become inefficient. (You would spend all your time digging small pieces of wire, nails, *etc.*) However, for archaeological purposes, the VLF and pulse induction types really come into their own. In such cases you would

Don't forget to search in and around all old outbuildings in your own neighborhood, even those that appear empty. I often hear of relics, especially guns and swords, that have been found in an old building.

want to locate and save ALL tiny pieces of metal, both ferrous and non-ferrous. Locating them before excavation could save a priceless artifact from damage. For this type of ghost-towning specialized instruments are certainly the answer.

The TR type will operate successfully if there are no large weeds or grass present. Many ghost towns in desert areas are rather clean and present no problems. Locations such as old cabin sites in remote mountainous or farming areas may have heavy vegetation and overgrowth. When this occurs in a mineralized area and the TR must be used, the instrument must be tuned in the NULL zone in order to eliminate erratic operation, in which case there will be a decided loss in sensitivity which is not acceptable. However, if you have no other type detector, go ahead and search with the TR.

The BFO type detector operates with continuous sound and does not have to be scrubbed on the ground. Thus, the BFO presents no operation problem for ghost-towning. Of course, the TR type, when it can be operated correctly, will detect coins easier and faster. Nevertheless, all locations, whether desert or mountain region, are different. You must take this fact into consideration.

Discriminators and Selectmatic types are not needed for ghost town hunting because a ghost-towner will want to recover ALL metallic items. The average detector, therefore, will perform quite well for the ghost-towner and the purchase of a discriminator is not absolutely necessary. It is well to note here that a few manufacturers produce the BFO, TR and VLF types in models that operate in the standard mode of operation but employ a switch that may be used to convert the instrument into a true discriminator. This convertible type of instrument, especially in a BFO, has many advantages for the ghost-towner since the BFO will respond audibly to all conductive targets, good or bad, while being employed for many all-purpose uses; and the advantage of metered discrimination, in addition to the already popular list of BFO accomplishments, is indeed marvelous.

Because your main goal will be to detect small, coin-sized objects and at the same time be able to gain enough depth to find shallow relics, the standard 8-inch searchcoil is the most logical choice. It is small enough to detect old coins but large enough also to detect larger relics at reasonable depths. Generally, the 8-inch coil is standard on most TR detectors. If your choice is the BFO all-purpose type, you might consider the independently operated dual coil. One of the dual coils manufactured features a 6½-inch coil and a 3½-inch coil. The small coil enables you to pinpoint the smaller coins. Small search-coils such as those used for coinshooting can detect coins but have less depth than the middle-sized loops, especially when detecting larger objects. The large search-

SEARCHING FOR GHOST TOWN RELICS AND ANTIQUES. The discovery of a buried treasure cache or a relic occurs every day at some remote ghost town. Almost any type detector will prove reasonably successful.

coils used to search for deeper treasure and money caches would locate the relics but fail to detect small single coins. A few TR detectors are equipped with only one large (10 to 12-inch diameter) searchcoil. These are popular in some non-mineralized areas, but due to the large size of the searchcoil they are highly erratic and almost impossible to use in most mineralized areas. If ghost-towning is intended to be your sole hobby it is possible to purchase high quality detectors at reasonable prices when the model carries only the middle-sized, utility-type (8-inch) searchcoil. If you wish to extend your activities to look for deeper caches or to prospect, choose an all-purpose detector. No matter what type detector you own or consider buying, the ghost towns will provide an endless variety of interesting metallic objects that will contribute greatly to the enjoyment of your hobby.

This very old ornate, intricately carved knife was found in West Texas.

I am indebted to Stanley Frank of Natchez, Louisiana, for this picture. The pistol is a rare Smith and Wesson .44 revolver that was developed for Russia. If you are ever in Louisiana do not fail to go by to see Stanley and his museum. He has one of the most fantastic collections of Civil War battlefield weaponry that I have ever seen . . . truly almost unbelievable.

With your detector chosen, all you have to do is locate a site that may have accommodated a number of people. Spot-check several areas to see if anything is there. It does not take much to confirm a possibly productive location — perhaps an old nail or some other meaningful object — and doing so will insure that you are in the right area. If you have sufficient time sweep the entire area with your searchcoil. If not, choose the most likely-looking spots and pay close attention to all response signals, both metal and mineral. Sometimes an old relic will deteriorate to the point where it may respond as mineral due to its extremely rusty condition. (It is returning to the natural state from which it came — magnetic iron — leaving only a trace of iron oxide to mark the spot.)

Tommy T. Long using a VLF type detector and Allan Cannon using a BFO type are searching for hidden crevices that could possibly contain hidden loot or small metallic relics. Located in a lava formation, the cave was used by Indians and outlaws for shelter. While road construction recently exposed and destroyed much of the cave, many relics have been recovered in this area. The VLF type detector operated perfectly here and the BFO was only slightly affected by the mineralization content of the lava formation. TR type instrumentation would have been almost impossible to operate because of the irregular, jagged surface and mineralized rock. Giving careful attention to operating characteristics of the different types of detectors can often spell the difference between the success and failure of a project.

There are many books on the subject of where to search. All will help to familiarize you with different areas. It would take volumes merely to touch the surface of this exciting hobby, but you will find that common sense and a good detector will generally take care of all your problems. Few, if any, ghost-towners return empty-handed. Perhaps your vacations take you to many unusual and interesting areas, possibly ghost towns and abandoned mining areas. The relics and old coins are there just waiting to be found.

Do not pass up the opportunity to locate barbed wire with your detector. Any type of detector can be used to locate buried coiled or single strands of all types of wire. The detector response may be either positive or negative, depending upon the type of detector you are using, the type of wire, how long it has been buried, and whether it is a single strand or coiled. It is quite easy to trace buried single strands. This fact will greatly aid you when digging the wire.

Check with a detector dealer or manufacturer. It is likely they stock many interesting ghost town maps or books to help you to pinpoint exact locations. Of course, the best source of information is a library. Its use costs you nothing, and in a few hours a lead can be quickly researched. This is possible even when on vacation or traveling. The librarian will guide you in selecting reference books. You may depend on this free source of information as being more accurate than hand-drawn maps or guides. Countless thousands of people on vacation travel through the most interesting areas without ever stopping! If they do stop they generally receive little help or information from questions put to the local inhabitants and become discouraged. Remember, the next time you wish information regarding ANY area go to the local library and ASK. This is the source used by most professional treasure hunters.

There is no end of places to search. Leave an area as

POST HOLE BANK RECOVERY. Many "family bank" recoveries such as this one have been made with the money hunter's companion, the modern metal detector. Due to the many different types of detectors (especially the new VLF types) being developed through today's advanced technology, many small caches of this kind can now be easily detected.

you found it — clean — and be sure to refill all holes. Use courtesy at all times. You might want to go back.

CHAPTER 9

Locating Antique Bottle Dumps

More and more bottle collectors are turning to the metal detector as an aid in searching for old bottle dumps. As the hobby of antique bottle collecting has gained in popularity the last few years more and more digging spots have been cleaned out. The only thing left to do is to use some sort of tool or electronic device to locate the tin cans and other metallic items that generally abound in dump areas. After much experimenting with the bottle probe and other underground locating devices, I have found the metal detector to be the fastest and the most practical instrument to use.

You may choose any type of metal detector, TR, VLF, RF (two-box), pulse induction, PRG or BFO. Of course, the PRG will require mineral-free ground; the VLF and pulse induction types are super-sensitive to *small* pieces of ferrous iron which presents some problems. The large searchcoils for the TR are extremely heavy, plus erratic in mineralized areas. The BFO requires large searchcoils to gain depth. However, all types of detectors will produce results with varying degrees of success.

Your choice of detector is not important, but you *should* use the larger searchcoils to gain extra depth. A few old dumps may be relatively shallow, and the standard coils will be adequate, but most of the *really old dumps* are rather deep. In this case larger searchcoils are necessary (the larger, the better) as the dump is usually a mass of metal and the speed with which you can cover unknown search areas will be of prime importance. If you are serious in your searching it will be best to choose fully shielded searchcoils. These coils eliminate grass and weed interference that is usually encountered around bottle dumps. However, as has been stated, you will have some degree of success no matter what detector type or method you employ.

With the larger searchcoils make your swing or sweeping motion in a circular or straight pattern, depending on your preference, at a height of about four to six inches above the ground. This will free the large coil from some of the mineralization interference and permit you to hear a faint response with minimum interference. You will find a ground probe helpful to save needless digging. When a signal is received, insert the probe before you dig to try to define what and how deep the metal is. Sometimes there will be NO glass in a small trash dump. By using the bottle probe in conjunction with the metal detector you will save endless hours of unnecessary digging.

Carl Young of Arlington, Virginia, was generous to share this photo of valuables he has found with his detector. A BFO type with a full complement of searchcoils allows wide diversification in treasure hunting activities — from coin hunting to relic seeking and bottle dump locating. The coins and jewelry displayed are some of his better coin hunting finds. The bottles came from an old dump he discovered.

Always operate your detector in the metallic mode of tuning. Keep in mind that even bottle collectors (whose prime interest is to secure antique glass specimens) sometimes discover a treasure cache. The large coils give fantastic depth, and you may just recover some deep items that other searchers missed with smaller coils.

Many ghost towns and other areas that have produced good bottles are constantly being reworked by collectors using metal detectors. Locations that have been searched time after time are still producing. As the average collector gains knowledge from books on bottle collecting and from information on new detectors, he is becoming aware of the fact that he just may have left some real goodies. The metal detector will never replace the shovel and bottle rake, but it certainly can find concealed dumps others have missed. Since the demand for rare, embossed whiskey bottles remains high, consider the worked-out areas that previously produced good finds. Common antique bottles have dropped somewhat in price, but the rare ones have increased at an unbelievable rate. This has always been true of priceless antiques, regardless of depressions or inflation. During the recent bottle boom I dug, bought and sold over 75,000 antique bottles on an international scale in a period of ten years. Over fifteen percent were found or recovered by use of a metal detector.

Never dismiss or leave a good location just because it has evidence of prior digging and previous searches. It is almost impossible for any hunter to clean a site completely. Consider well the advantage of using the metal detector before you abandon the search of an area.

BOTTLE DIGGERS continually find single old coins and occasionally a can or jar full of coins that was discarded by mistake or buried and forgotten. Larger items . . . chairs, boxes, light fixtures or any enclosed container, for example . . . should be carefully inspected because small amounts of coins may have been secreted there.

Bill Mahan, Garland, Texas, is often called "King of Padre Island." Bill has probably found more relics and money on this historic island than all other hunters combined. He most surely opened the island to exploration by his recoveries. It is now a National Seashore and closed to all searching. A great gap will be left in the known history of ships wrecked during the days of the Spanish explorers. Bill, one of my dear friends and an old hunting companion, is a true professional who ranks as one of the great treasure hunters of all time. If Bill gets to the Happy Hunting Ground first, I know I will find many deep and empty (but filled in) holes waiting. Many thanks to Bill and his wife, Joyce, for this picture of just a few of his fabulous finds.

CHAPTER 10

Building and Cabin Searching

Searching an aboveground structure in an area free from the presence of ground minerals is especially easy. Any amateur or beginning treasure hunter can tune a metal detector inside a building constructed of wood or other non-conductive material and tell instantly where coins, a can, or a metal box may be hidden in a wall.

Almost any type of detector will work perfectly under these circumstances, PROVIDED the structure does not have nails and other small ferrous targets present. Many dealers will demonstrate their detectors inside their shops, perhaps on the wooden floor. However, this procedure can be somewhat confusing to the prospective buyer because no mineral is present as there may be in an actual field operation. Some confusion may also be generated as to which type of detector operates best when used among nails or other small ferrous junk. Since most dwellings, even hundred-year-old log cabins, have some ferrous material present, correct selection of an instrument is critical.

A cache that is still intact will generally be in some type of ferrous or metal container. This fact automatically rules out the use of ANY type of discriminating detector. You would wish to receive ALL metallic indications and, if possible, try to evaluate them as to size and location. Such evaluation can be very difficult if nails and other small conductive items are imbedded in the walls, ceilings and floors. Different detector types react differently to small ferrous items, not only to the type of item but also in relation to the way in which the item faces the detector searchcoil. The following paragraphs provide a reasonable description of the way each type will react under average conditions, in average surroundings. The information is in no way meant to suggest that ALL conditions

I don't particularly advise crawling under old houses and buildings, for obvious reasons. However, a well-trained detector operator does not need to if he goes about his business in the right way. Here Bob Podhrasky scans for metal objects that may have been concealed in the walls or placed on the beams of the old house. An entire house can be scanned in about fifteen minutes. Due to the presence of numerous nails the BFO is always mandatory in this type searching.

or EVERY building will produce the same results. Only the natural reactions of detector types under average conditions are described.

PRG TYPE

The sensitivity, stability and target identification of the PRG is beyond question equal to or better than that of any detector currently on the open market. However, any quick response, super-sensitivity type instrument, *regardless of whether it discriminates*, is almost helpless when used where small ferrous targets exist. The PRG type is also slightly heavy (almost seven pounds) for efficient overhead or tight space maneuverability. It is somewhat expensive to purchase to use for a relatively

Always scan above doors and windows and around the door and window sills. It is reported that one treasure hunter found a $50 gold piece valued at several thousand dollars when he raised a wooden frame window in an old house. The gold piece had rested beneath the window frame for about 70 years. Always scan the baseboards, especially any that look as if they could be trap doors or easily removed.

simple operation such as building searching. The discrimination readout does little or no good in building searches. The PRG performs almost perfectly on or near salt water beaches relatively barren of mineral content and should be reserved for use where it performs best.

PI TYPE

The pulse induction detector is super-sensitive to small ferrous targets (such as nails). Since there is no need for the capabilities of its almost mineral-free operation, it is not suggested for use where nails or other small ferrous targets exist. (It would be almost impossible to pinpoint a metal target among the presence of many nails.) It should be employed where its advantages are obvious. Experience with this type instrument is required for the operator to learn to discriminate with reasonable accuracy, but discrimination in building searching is not desirable. Use your PI where it performs best . . . in security matters and searches conducted by trained personnel, plus beachcombing for coins where ferrous trash is not abundant.

TR TYPE

The standard or conventional high-frequency transmitter-receiver type is practical for building searching under certain circumstances. Discrimination adds nothing to its capabilities since you want to find any type of small ferrous container (with the money INSIDE). Also, small ferrous targets will produce different indications, depending on their position in relation to the searchcoil. A

When searching old houses, be certain to investigate any places where flooring has been repaired, patched or otherwise altered. In this picture there are four or more such places which should be investigated.

Ah, here's a favorite place! If you have ever heard the old man brag about how sweet his well water is . . . then you should search in and around the well. It just may be that he was unable to retrieve his cache before he died. Search the path between the well and other land features such as buildings and trees. Often a ledge was formed a few feet down into the well which permitted valuables to be secretly cached out of sight. Always scan every suspicious-looking place that could be either a trap door or portion of flooring or walls that could be removed to conceal valuables. In this photograph there are at least four such places.

nail pointing toward the searchcoil may produce either a positive or negative response, depending on the configuration of the searchcoil. The TR type is a quick response instrument, almost always responding wide open (full audio signal) so there is no way to determine the size of the target. A TR may be productive under many conditions and, of course, will continue, for some time, to be used as a coin finding detector. However, regardless of its popularity in that field, it leaves much to be desired when attempting to read, among all the nails, what is behind a log or wall. The response of any type TR is simply too quick to be practical for building searching and requires the operator to investigate all tiny indications. If it is a "have to" case, it could be used for building searching. There will always be, however, much room for doubt as to whether you missed a small cache among all the other small indications, especially if you must detune the TR for use among the other metal. It simply does not perform efficiently where detuning is mandatory for ease of operation. You may miss the boat!

VLF TYPE

For building searching the VLF is probably the most practical, certainly the most sensitive, of all detectors in the TR family of instruments. However, it is almost impossible to use to conduct a building search where small

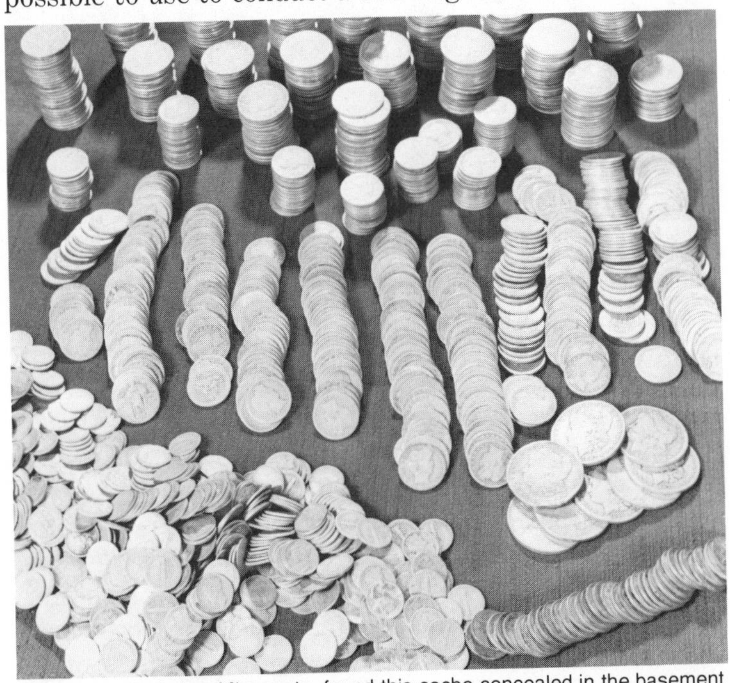

Bill Mason, Redwing, Minnesota, found this cache concealed in the basement wall of an old house. Bill employed a BFO type detector because of the nails and other small metallic objects that are almost always present in, under and around old buildings. Bill is one of the more experienced cache hunters in the business, and he and the author have made many trips together on treasure leads. This particular cache was exhibited by Bill at the 1974 Oklahoma City Treasure Show hosted by Bob Barnes at Shepherd Mall. Thanks, Bill, for showing our readers a real treasure.

nails are present. While, with its super-sensitivity and mineral-free operation, it is certain to replace most other detector types, use of the VLF among small ferrous targets is inefficient, whether or not the instrument discriminates. You are looking for ferrous containers and wish to miss none of them. The VLF is super-sensitive and you would not want to *detune* it, for that results only in the possibility of missing a cache by FAILING TO HEAR THE RESPONSE. This super-sensitive instrument is truly the detector of the future, but under certain conditions super-sensitivity can be a hindrance rather than a help. Always utilize any detector type in the area where its performance is rated best.

Bob Podhrasky, Garland, Texas, is using the old reliable workhorse of the detector industry, a BFO type instrument. Bob is searching for a small tobacco can reputed to be hidden in this old attic. Notice the presence of numerous nails. The BFO is a "must" under these adverse conditions. Bob is one of the most highly regarded electronic engineers in advanced circuitry design, having adapted the P2L (phase locked loop) method tuning into the first truly stable system of instant retuning. This method has been adopted by most manufacturers in the business. Many thanks, Bob, for your engineering skills and your willingness to share the fruits of your labor with the man in the field.

Tommy T. Long, hunting and business partner of the author for many years, found this pistol in an old building in Colorado. When Tommy T. travels, he always carries two kinds of detectors, a BFO and one of the very low frequency type. The latter is for seeking extra-deep caches in mineralized ground. "It makes no sense," says Tommy T., "to use a small coin hunting coil for cache hunting. Most caches are far beyond the depth detection limits of those small coils." If more treasure hunters would learn that, much more treasure would be found. The right equipment for the job must always be used.

BFO TYPE

This detector type does not see most of the nails used in wall construction, especially a small-headed finishing nail. Even when used among other metal the BFO will continue to gain in beats, enabling the operator to gauge the size and distance of the target. The BFO operates with a continuous beating or humming tone which prevents missing even the slightest indication that might result from among the rubble of small ferrous material. The BFO is the slowest response detector among all the types, but this characteristic can be very advantageous under certain circumstances. Efficient operation among assorted metals or minerals is one of the advantages of the BFO type. With experience and patience it is possible to learn to gauge the size and depth of a target.

Discrimination will, again, be of no advantage as you are generally trying to ferret out a metal container. This has even been accomplished with a BFO type in old fireplaces and ruins where other small metal was present. The operator was experienced and could distinguish the size and presence of larger metal targets among the smaller indications. A tremendous advantage of the slower response characteristic is that the operator can continue without detuning the detector. The BFO with small coil attachments can be maneuvered slowly in tight or small enclosures and the slightest increase in beats or tone is easily heard. It does take more time to master the BFO type for use in certain other situations, but building searches can be conducted quickly and easily by even a young child. In this type of searching operation the BFO will ALWAYS maintain its superiority due to the continuous sound factor and the slower response. I suggest you try the detector best suited for this type searching.

CHAPTER 11

Coin Hunting

Single coin hunting is undoubtedly the most popular phase of the treasure hunting or hobby fraternity.

The search for single coins is relatively simple and easy and need not damage the grass or ground from which coins are removed. Anyone — small children to adults — can participate. All it takes is a good quality metal detector and the urge to search!

Coinshooting is gaining in popularity around the world as hobbyists in Greece, Vietnam, Indonesia, Europe, Australia, and many other countries join in this exciting and profitable pastime. Interest and enthusiasm is further generated by coin hunting contests and fellowship meetings of the numerous treasure and hobby clubs. Those who wish to find out about such clubs and their locations will find the following magazines among the best sources of information: *TREASURE*, *TREASURE SEARCH*, *TREASURE FOUND* (Jess Publishing Company, Inc., 9420D Activity Road, San Diego, California 92126); *WESTERN AND EASTERN TREASURES* (People's Publishing Company, 1440 West Walnut Street, Compton, California 90220); and *LOST TREASURE* (Carter/Latham Publishing Co., Inc., P. O. Box 328, Conroe, Texas 77301). They are available on most newsstands or by subscription. These magazines are some of America's leading publications on treasure hunting and recreational mining. They are a "must" for any outdoor enthusiast.

CHOOSE YOUR DETECTOR

Your choice of a metal detector to enter this exciting

Success sometimes comes in small packages. For years I have been stressing that the treasure hunter should follow the bulldozer and house wrecking crews. Here is another story where the bulldozer did its job. All of these rare coins were found by one successful treasure hunter near an old home site in Dalles, Oregon, after a bulldozer cleaning the site unearthed a fruit jar containing these particular coins.

hobby may include all different types. The transmitter-receiver type detectors (includes VLF's and TR's) are the best for this hobby. Possibly this is due to the almost silent operation or the quick response feature which causes them to respond so well to single coins. Nevertheless, almost all types are used extensively in this hobby. Consider the field trials that have been conducted across the nation. You will notice that the various types have performed equally well. It is up to the individual as to which he likes best for his particular purposes.

Many thanks to E. S. "Rocky" LeGaye for sending this picture. Karen Pafford, a Texan, has become quite excited about treasure hunting since she found her long-lost ring. She has become so enthusiastic about the treasure hunting hobby that she is now planning to search for a cache known to have been hidden somewhere on her family's old farm in East Texas. This detector is one of the earlier model, single coil type discriminators. Most single coil discriminators have been replaced by more modern, independently-operated dual coil detectors which are capable of cache hunting and discrimination while coin hunting. Photo by Howard Robenstein, Channelview, Texas, who writes for the Houston Sportsman's Club paper, *News & Views.*

SINGLE COIN HUNTING is without question the most popular phase of treasure hunting. The VLF and TR types are the most popular for this particular hobby.

With a TR detector the standard 8-inch coil is certainly the most practical. With the BFO the 3-inch, 5-inch, and 6-inch coils will be the basic choices. The discriminators or Selectmatics usually come equipped with only one small searchcoil that is attached directly to the circuit board. Most of these coils cannot be changed or removed. The PI (pulse induction) responds well with the 8-inch coil. The PRG may use the 8-inch coil in non-mineralized areas, but the smaller searchcoils work better in areas where light mineralization is present. The RF (two-box) does not respond to single coins. The VLF types are generally designed with either 6 or 8-inch searchcoils. The 8-inch is the most practical and best producer on small targets, provided it is designed to pinpoint accurately. Faraday-shielded searchcoils are very important on any type of detector. Interference caused by weeds and grass on unshielded coils will sometimes so distort faint signals from deeply buried coins that the signals are impossible to hear.

Detector sensitivity, Faraday-shielding and correct searchcoil size are the prime detector requirements for coin hunting. (Of course, reasonable attention must be given to an appropriate area of search.) In this chapter we will deal primarily with the selection of your detector for coin hunting and how to operate it properly to the fullest advantage.

THE TR

If you have chosen a TR detector, set the tuning in the metallic mode of operation. Place the searchcoil flat on the surface of the grass. Adjust the tuning slightly until

Mr. Chris Adams of Rochester, New York, displays some of the silver coins he has found while coin hunting. Note that all these coins are silver which means that you can multiply their face value by a factor of at least three. An added benefit to coin hunting is that when you find old silver coins such as these, they are worth much more than face value.

you hear a faint background of response and move the searchcoil in a sidewise sweep or circular motion. This will enable you to overlap each sweep slightly and prevent the missing of any coins. Some operators like to tune the TR in the null or dead area. While this adjustment does make operation more silent and perhaps less irritating, it also increases the chance of missing some of the deeper coins. If the TR is tuned silent there is no way to tell when mineral change or slight drift forces the tuning farther back into the null area. This has the effect of losing part of the detector's sensitivity.

It is up to the individual as to where he likes to set the tuning. Practice will always be the deciding factor. It is not mandatory to operate the TR searchcoil flat on the ground's surface. You may wish to operate one or two inches above the ground. You will, however, gain a couple of inches of depth with the ground scrubbing action of the coil and you will get tremendous depth on single coins if you keep the detector properly tuned and slightly up in the tuning range.

THE BFO

If you have chosen the all-purpose BFO detector, the operation differs little. The BFO searchcoil will probably operate best at approximately one to two inches above the surface of the ground, depending on ground mineral content. Set the tuning in the metallic mode of operation at a speed of anywhere from 20 to 60 beats per second. The speed setting may vary according to individual preference. Some like the moderate motorboating sound; some like faster tuning. The faster speed is *absolutely* necessary with BFO high frequency detectors or with poorly designed circuits that are highly erratic and unstable. You will find that the BFO's constant sound feature makes a signal harder to distinguish than the "whap" or quick response of the TR, but with practice you will be able to hear even the smallest increase in beats. Many hunters

Mr. George Banks holds the coin he found that is estimated to be valued at approximately $300,000. There are only two of these genuine coins known to be in existence. One is owned by the Ford Foundation. George kisses and blesses his faithful BFO Cache Hunter with which he found the coin. It was found with the small 5-inch coil of the independently-operated dual coil configuration. The ability to change rapidly at the flick of a switch from coin hunting to cache hunting has many times greatly assisted the detector operator.

even actually prefer the constant sound feature for single coins.

The depth that is achieved by the BFO may be somewhat less on small coins than that of a TR, but it is easier to use on rough ground. Fully shielded searchcoils are very important on ANY type detector, especially in the search for single coins. Deeper coins produce fainter signals, and grass and weed interference produced by unshielded searchcoils garble and destroy faint signals, making them barely distinguishable. BFO searchcoil choices will include the independently operated dual coils which are very popular because they allow you to conduct your search with the larger coil then switch to the smaller, inside coil to pinpoint a signal. This arrangement does have an advantage in the amount of search area which can be covered, and with proper practice you can become quite an expert with an independently operated dual coil.

The rate of your searchcoil sweep or swing should be governed by the tuning speed. If your detector is tuned at a slow motorboating sound, move the searchcoil slowly and keep it level with the ground. If your instrument is tuned at a moderate or fast motorboat sound, then you can move the searchcoil faster. Moving the coil TOO slowly will fail to achieve the "break" or increase in beats; TOO fast a pace will cause you to miss the target. More practice is required with the BFO for locating single coins, but experienced operators will hold their own in most situations and, *under certain adverse conditions,* will even surpass the performance obtained with standard TR's.

THE VLF

Perhaps you have chosen one of the specially designed VLF types. These detectors feature the GREATEST DEPTH ever obtained on coins or small metallic objects. However, without some type of discriminating ability these deepseekers are almost impossible to use in trashy or littered areas. They are extremely supersensitive and will detect tiny ferrous and non-ferrous

The author, Bob Grant, and Frank Mellish check out a rock sample that responded as conductive to the VLF. Often times miners dropped valuable ore specimens as material was being removed from the shaft. A good detector can indicate the predominant material as being conductive ferrous, non-conductive ferrous or conductive non-ferrous. Of course, all conductive non-ferrous material should be brought in for further testing.

objects so small the eye can hardly see them. They will detect nails at fantastic depths and are in constant demand by loggers, sawmills and surveying crews. They will operate perfectly in highly mineralized areas with NO LOSS IN DEPTH, finding deep coins that are impossible for other detector types. If the area you intend to search is relatively free of junk and has been hunted previously, these deep-seeking VLF types will produce the best results.

The mineralization may be completely zeroed or tuned out, permitting the searchcoil to be operated at any given height. This enables the operator to conduct his search over mineralized river gravel, in areas containing concentrations of black sand, and in other difficult places where the standard TR is almost impossible to operate and where the BFO type experiences a definite loss in depth. Simply adjust the searchcoil to the ground by use of the ground or terrain control according to the manufacturer's instructions. (Do not confuse this control with conventional "ground" or "mineralized" adjustment features on standard TR's.) You may then adjust the tuning to the desired level of sound and proceed to search.

The searchcoil may be elevated to any practical operating height. After you receive a hot or loud signal you may find it sometimes necessary to back off or detune the sensitivity in order to pinpoint the target. You can now understand the problems that you may encounter in coin hunting with super-sensitive instruments in trashy or littered areas. The middle-size (8-inch) searchcoils produce the best results. Larger coils would only make the problem of pinpointing more difficult.

Different manufacturers use different configurations in their coil designs. Most TR (IB) searchcoils respond to metal on both the top and bottom sides. Searchcoils designed with bottom detection only have a more even or uniform detection pattern when operated in close proximity to the coil, and this enhances their ability to pinpoint more accurately. This configuration design is commonly called a stacked coil, but the technical name is co-axial. It is used on many of the more sophisticated instruments and has none of the dead spots common to TR (IB) searchcoil designs. (Still, it will not produce true metal *vs.* mineral identification on small, marginal ore specimens.) The stacked coil design reverses its detection pattern as the target passes through the center of the coil. This has no relation to coin hunting, but the lack of interference from small metallic objects which may be on the operator's clothing is a definite asset, especially with these supersensitive instruments. Also, the co-axial searchcoil configuration almost completely eliminates outside electrical interference common on conventional co-planar types when operating in the very low frequencies.

Consider the very important ability to pinpoint when selecting your VLF type detector for coin hunting. While attempting to pinpoint your target you may also achieve a certain percentage of discrimination by passing the searchcoil forward and backward and side-to-side with a left-to-right and right-to-left motion. If the target is an elongated ferrous object such as a nail or small piece of iron longer than it is wide, you will probably receive a double "blip blip" response. If you receive only a single "blip" the target is either non-ferrous or perhaps a round iron object. This method of partial discrimination lacks total accuracy, but interpreted with operator experience it can save a lot of useless digging.

Drift is hardly ever a problem with these supersensitive instruments, but audio and meter response may be somewhat erratic due to the very low frequency operation. Close proximity of electric power lines, underground cables and power stations tend to distort the response on some detectors which employ the common or conventional TR searchcoil. This knowledge will assist you in selecting an instrument to provide added depth in old parks and playgrounds which may be adjacent to power supplies.

As Dick Blondis will agree, England is a treasure hunter's paradise. My thanks to Dick for this photograph of a portion of his collection of old Byzantine bronze reliquary crosses and medieval medallions. Today's new ground canceling, very low frequency types of detectors will permit operators to go back and rework highly mineralized areas such as the places these artifacts were found, making the sites productive once again. Photograph by Polner.

Since VLF types produce fantastic depths on older coins, many supposedly worked-out parks and beaches will again give up their treasures. Of course, a VLF with discriminating abilities would be a perfect choice for coin hunting by eliminating most small ferrous junk.

THE PI

If your choice was the pulse induction type detector the degree of sensitivity or method of operation will differ little. The PI is much lighter and features a searchcoil that is completely unaffected by grass, weeds or salt water. The response of the PI is a bell-like ringing which makes pinpointing difficult. You must detune or back off the amount of sensitivity to narrow the target area. The PI features earphones, but a small outside speaker is simple to add. There is no zero control to adjust for mineral content, but it does not need one. Merely turn the tuning control on and advance it forward until the point is reached where the bell starts to ring. Back it up slightly, and you are ready for operation.

Start your search by holding the searchcoil approximately two to four inches above the ground, taking care not to vary the operating height too greatly. This is important as the PI will respond positive (metallic) if brought into contact with mineralized ground. This presents no problem if you keep the searchcoil height approximately level. The main drawback is pinpointing the indicated target in relation to or among other small metallic objects. The instrument's super-sensitivity to small ferrous (iron) targets sometimes makes operation impossible. Expert operators achieve a certain amount of discrimination with this detector, but its scope of operation is best limited to certain areas for coin hunting.

THE PRG

The PRG type produces almost perfect target identification and the sensitivity compares with the VLF and pulse induction types. It operates perfectly in salt water areas as it is unaffected by mineral salts. Most detector types are affected by mineral salts since when the salts are WET they become conductive and respond as metal. The PRG does not have this fault, and if used in selected zones of operation can be very successfully used for coin hunting. It will NOT, however, operate adequately in mineralized zones. It is adversely affected by even a small amount of magnetic black sand (Fe_3O_4). As any type of discriminating detector sees magnetic black sand as BAD, a problem is presented for a sensor or circuit that has sensitive target identification. The greater the target identification any discriminating type has, the more difficult the instrument is to operate over ground with magnetic mineral (bad) content. The PRG is rather heavy, about seven pounds, and requires some operator skill in operation. It might be considered expensive, but not for selected coin hunting in specific locations, such as in salt water areas where valuable Spanish coins may be found. If you are coin hunting in zones free of magnetic content the PRG will produce almost perfect identification on targets and achieve fantastic depths. For best results follow the manufacturer's instructions on tuning and operation.

THE DISCRIMINATORS

Perhaps you have chosen one of the many different types of discriminating detectors, commonly called "discriminators," "Selectmatics," "Rejectors," *etc.* These detectors are equipped with either a meter or an audio conductivity readout indicator. This means the detector will discriminate between non-conductive ferrous iron or metallic targets of LOW conductivity and targets of HIGH conductivity, rejecting bottlecaps, small pieces of ferrous iron, nails and tinfoil. This capability can be quite helpful when coin hunting in trashy areas, but, since many metallic items contain almost the same conductivity factor as some coins, it is practically impossible to obtain 100%-accuracy.

Pulltabs from beverage cans are among the marginal targets. Some manufacturers have adjusted the conductivity indicator to reject pulltabs but with a considerable LOSS in sensitivity. Others have compromised and accepted the marginal target as good, attempting to retain all possible sensitivity. A self-conducted sensitivity test on different models can prove very enlightening.

It is virtually impossible for a discriminating detector to retain all the desirable qualities of the quick-response TR (IB) and the all-purpose capabilities of the BFO. However, if your principal hobby is coin hunting in parks and restricted areas where trash abounds, you will find a discriminator a joy to use. The slight loss in depth will be compensated for by eliminating 50-90% of the unnecessary digging. The perfect solution would be to acquire both a conventional detector and a discriminator or at least two discriminating detectors, one with both a normal *and* discriminating mode of operation. This should not be too difficult to accomplish when an entire family is involved in treasure hunting with metal/mineral detectors. As there are many different types of discriminators with greatly differing designs, correct field operation is best accomplished by following the manufacturer's recommended tuning procedures.

A few of the larger manufacturers produce BFO and TR types that employ a conversion switch or control to change instantly from normal mode to discriminating mode. This is extremely helpful as you may conduct everyday treasure or prospecting searches with full sensitivity, plus metal *vs.* mineral identification when using the BFO type. If discrimination is desired, just SWITCH to the discriminating mode. Some of the BFO types have a variable discrimination control that increases the amount of rejection WITHOUT LOSS IN SENSITIVITY. Many TR types are designed with mineralization controls that aid the operator when adverse soil conditions are encountered. This, in effect, also reduces the amount of sensitivity, but it is sometimes necessary.

All of these advanced features are very helpful. The slight increase in cost for a detector that features both discriminating and normal modes of operation can be easily justified by increased ease of operation and productivity. This dual-circuit type of operation occasionally demands a special type of searchcoil in TR models which can sometimes make the discriminating coil more difficult to operate over mineralized ground. It is wise to purchase through a local dealer so that you may observe actual operation. In addition, you should ALWAYS consider the reputation of the manufacturer in regard to detector quality. There are many types of dual-circuits, and correct field operation is best accomplished by following the manufacturer's recommended tuning procedures.

When making a choice among a standard TR, VLF, PRG, IB, PI, BFO, single coil distriminator or twin circuit type, evaluate all the evidence. The choice should depend on the individual buyer. ALL detectors will find coins with some degree of success, depending on operator skill. Some will locate coins at reasonable depths; some, at great depths. A few specially built detectors operating in the extreme low frequency range with high gain will recover coins at depths you will find hard to believe. Of course, the type of discrimination employed on these VLF type detectors will differ greatly and will cause varying depth capabilities.

A *TRUE* mineral-free discrimination method, if available, would suffer no loss in depth. This method of total, mineral-free discrimination could safely be considered the ultimate goal of the coin hunter. Of course, any method of discrimination that would involve the switching or changing over from the VLF's true mineral-free mode into a standard TR discriminating mode would cause some loss in depth if the detector were operated over highly negative or mineralized ground. This is caused by the fact the very low frequency TR sees the mineralized ground more quickly and is MUCH MORE adversely affected by it than is the standard high frequency TR type. This method of discrimination requires the gain or sensitivity to be reduced when the detector is operated in mineralized areas, losing some of the advantages of the VLF's deep seeking capabilities.

To summarize, I recommend that all coin hunters consider their locale and surroundings. Those who search parks and areas that have been previously hunted should choose detectors that produce GREAT DEPTH to recover the deeper and possibly more valuable coins that have been missed. Trashy, junk-littered areas will best be searched with a type of detector that has discriminating

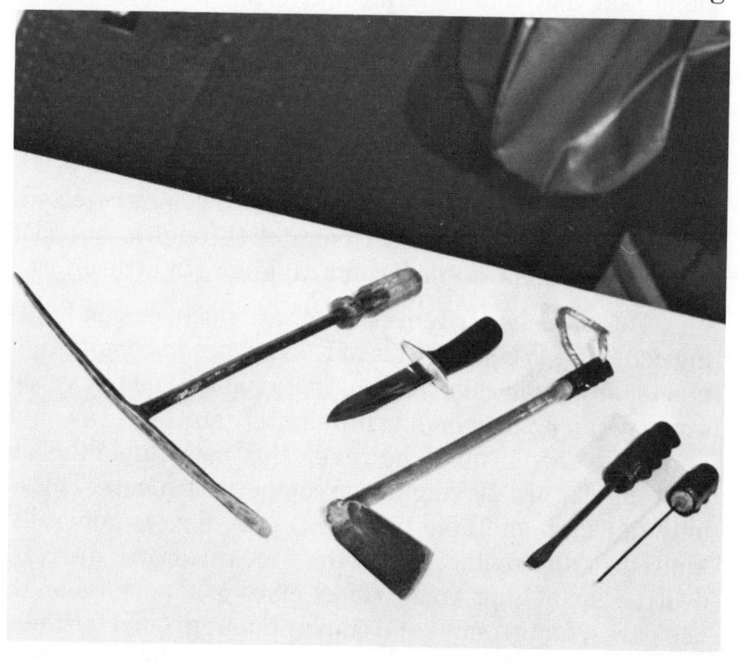

The combination rake and hoe tool is made from a small garden scratcher. It is meant to be used *only* where slight digging or scratching does not damage or destroy the gound's surface. This picture shows many variations of diggers. You will have no trouble constructing one of your own to save hours of backbreaking labor in places where use of such a tool would be permissible. The screwdriver and probe (right) are probably two of the most useful and essential tools employed in coin hunting, especially in areas where care must be taken with the turf.

abilities. Content of soil mineralization should always be taken into consideration. Some detectors will operate perfectly in highly mineralized areas; some, only adequately; and some will not perform at all. The ability to penetrate mineralized soil with ease of operation should always be one of the FIRST considerations when selecting a detector for coin hunting.

RECOMMENDED DIGGING TOOLS

Many different types of tools can be used to recover single coins speedily. Everyone has his favorite invention. A *combination rake and hoe* is made from a small garden scratcher. Attach a longer, sturdier handle on any type of small rake and you have the equivalent of this tool. It is meant to be used ONLY where slight digging or scratching does not damage or destroy the ground surface. It is a very popular tool for searching around old homesteads and isolated areas. With it you can quickly scratch or dig a small coin or metallic object to the surface without bending over. If the item is unwanted junk it can be rejected. By using the end of the rake to scratch through loose soil it is possible to expose small coins rapidly with little effort.

This and similar tools are very popular in coin hunting contests. When speed is of the essence in competition meets, any small compact tool that enables you to expose a coin without backbending is definitely an asset. It would be wise to keep in mind, however, that most hunt officials limit the size of a digging tool in competition meets. This is only fair and equitable to all entrants. Size is generally limited to approximately one-half-inch width to prevent destruction of hunt areas. (Most clubs gain permission to use parks, fair grounds and similar public ground for their events.) Also it is mandatory that all holes be filled. Common courtesy and respect for other people's property have helped the coin hunting hobby grow throughout the nation.

The *combination screwdriver-ice pick* is probably

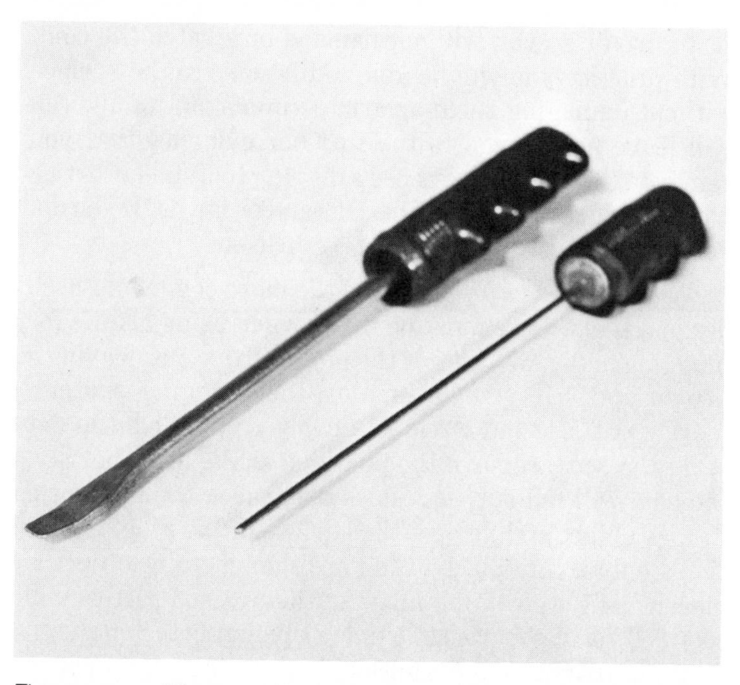

There are many different combinations of tools used for the removal of coins from lawns. Many hunters simply use a long, narrow screwdriver for probing and for locating the coin and popping it out to the surface. Some carry both the slender probe and screwdriver. Some, however, eliminate the need for two different tools by combining the probe and screwdriver into one unit. It is a simple task to drill a small hole in the handle of the screwdriver and insert the probe. Take care that the probe is securely fastened; otherwise, when used in hard ground it may become stuck and slip from the handle.

one of the most useful and essential tools used in coin hunting, especially in areas where care must be taken with turf. By careful probing with the ice pick it is possible to locate the coin, insert the screwdriver carefully under it, and lift it out. This tool can be built quickly and inexpensively. Bend the end of an average-sized screwdriver somewhat so you will have a slight hook to insert under the coin. Remove an ordinary ice pick from the wooden handle, drill a small hole in the end of the screwdriver (opposite the screwdriver end) just slightly larger than the ice pick, fill the hole with some type of strong glue, and insert the ice pick into the hole. When the glue is thoroughly dry, smooth the ends of the ice pick and the screwdriver so you will not damage or scratch the coin. With practice you will be able to lift coins to the surface without damaging them and, most important of all, you will leave no damage to the turf nor evidence that you removed the coin. Always use a digging tool similar to this when conducting a search in parks and on lawns. If you do, you will always find the welcome mat out.

The *hunting knife* is one of the more commonly-used digging tools. Care must be taken when using a knife for plugging in dry weather or the plug will dry out, leaving a brown spot. If you do employ the plugging system, merely cut a round circle as deeply as possible (do not scalp the sod), remove the plug and shake it. Generally, the coin will fall out. If it does not, check the hole for a deeper object and inspect the surface of the plug. Sometimes only a tiny piece of foil or other metallic object is merely caught in the top part of the grass. Practice will soon tell you if you wish to employ a plugging technique on turf. Sometimes it is permissible; more likely you will be evicted from the park or lawn area. Be certain to replace the plug and tamp it down firmly.

YOU CAN BE SUCCESSFUL

If you have purchased a quality detector, gathered your digging tools, chosen a likely looking lawn where the house is relatively old, and proceeded in the normal method of searching . . . YOU ARE ALMOST CERTAIN TO FIND COINS. This is why the coin hunting hobby is the largest in the world — anyone can be successful at any time, at any place.

If you are an avid fisherman you do not have to become a treasure hunter just to search for coins. Simply purchase the detector of your choice and carry it with you as you enjoy your primary outdoor sport. You will find that sportsmen everywhere are adding the interesting metal detector to their list of additional accessories, reaping happiness and enjoyment from filling in the extra

Cache hunter using the biggest gun of all . . . a large coil on a VLF type detector. This combination has the advantage of detecting *both* extremely small and large targets simultaneously while penetrating the most highly mineralized (iron, Fe_30_4) soil or rocks with ease. This characteristic may sometimes be a handicap in trashy areas, such as around an old homestead.

Cache hunter using a true, deep-seeking coil on a BFO type detector. These larger coils are absolutely necessary for detection of small caches at a depth of more than a foot or so. These larger BFO coils do not operate completely mineral-free, but they do have the advantage of by-passing some of the smaller ferrous trash (nails, bottlecaps, *etc.*) around old homesteads and other littered areas — a great aid when searching for larger-than-silver-dollar-size caches in such areas.

hours of leisure when "they aren't biting." This is true of the hunter, snowmobiler, back-packer, cowboy and golfer. Just think of the places you have seen when you have been following your primary sport or hobby. Think of the hours of interest a metal/mineral detector will create for your friends and family. When you have an extra hour to fill, remember that all over the world coin hunting produces more excitement and fun than any other hobby.

For those who wish to pursue this hobby to its fullest, I recommend Charles Garrett's book, *SUCCESSFUL COIN HUNTING*, published by Ram Publishing Company. It covers all phases of coin hunting with all types of detectors and lists hundreds of interesting places to hunt. It is the most informative book of its kind; no coin hunter can afford to be without it.

CHAPTER 12

Cache Hunting

As mentioned previously, the words "treasure hunting" actually cover ALL phases of detector usage. In this chapter, however, we will allude only to actual cache hunting or, rather, the searching for buried treasure, NOT for coins or small objects. Detectors vary greatly as they are pushed to the utmost in depth penetration for the larger targets. They will differ also as to the amount of response or signal received on these deeper targets. The mineral content of the soil will sometimes determine the response AND sometimes the size of the cache may determine the outcome of the search. Many beginners have hit the treasure trail in a desperate attempt to make it pay as some of the full-time treasure hunters have done. All have quickly discovered that it takes more than just a detector and ambition. Probably more myths and misinformation exist in the cache hunting field than in any other phase of treasure seeking.

So many factors enter into the successful detection of deeply buried caches that few people have achieved enough success to become experienced. A few of these factors are: the location of the suspected treasure site; the amount of vegetation present; the amount of adverse ground mineral content; the condition of the ground surface; how small or large the cache is reputed to be (a quantity generally overestimated); how deep the cache is buried; whether the ground has been built up over the cache; whether the detector you use can penetrate the type of soil or rock that is present; whether you have the correct size searchcoil necessary to obtain all possible depth but still small enough to respond to the estimated target size; *etc*. Do not complete your research and arrive at the suspected location before considering these matters. Since I know of no guide book or manual on the market that attempts to explain all these various factors, I will attempt to touch on every aspect of cache hunting, with every type of detector, in the hope of explaining why a past venture failed and to help you achieve success when you arrive at your next location.

If you decide to hunt elephants, use an elephant gun! A major failure of beginners is that they expect to find deeply buried targets with a small searchcoil or with the same detector and coil that was so successful in the search for shallow relics and small coins. You simply MUST employ large searchcoils for the recovery of larger, more deeply buried treasures. Regardless of over-enthusiastic advertising and tall tales told by inexperienced treasure hunters, you cannot change electronic facts. If an average

Money was cached in each of these three containers. The early model RF type detector shown in the picture was used to find these caches. Cache hunters are rapidly replacing their RF two-box detectors with the very low frequency types of detectors and the largest searchcoil they can buy. The very low frequency type of detector does not have problems with ground mineralization and detection of voids in water pockets as do the RF two-box type detectors.

size searchcoil, say, 8-inches, produces an electromagnetic field reaching three feet or even six feet, this DOES NOT MEAN it is capable of detecting an average size chest or kettle at this depth. It means only that you would get a faint indication on a large metallic target, such as a car body in mineral-free ground.

The length of time the target has been buried also enters into consideration. If you were to freshly bury a mass of metal at these great depths, the depth expectation could be cut in half, depending on the amount of mineral present in the soil and type of detector used. Bury a quart fruit jar full of silver or other metal from ten to twenty inches deep, a large kettle or Dutch oven from eighteen to thirty-six inches deep. Use the best detectors available with average size searchcoils and you will understand why most treasures are simply walked over and *missed* by hard-working treasure hunters who placed their faith in small or medium size searchcoils. Even the best deep-seeking detectors utilizing large searchcoils will experience some difficulty in this test.

The average cache is almost always as large as a small tobacco can and may extend to any larger size, such as a chest, kettle or even bigger object. This fact allows a wider choice as to type of detector used, but you will still need to employ the large searchcoils as insurance for sufficient depth. Listed in mixed order are the detector types and established guidelines for successful recovery of deep targets.

No slight is intended to any manufacturer who does not produce detectors capable of utilizing such large searchcoils. Many manufacturers produce quality lines intended only for the sole enjoyment of coin hunters. There are some low-quality instruments capable of using larger searchcoils, but they are so unstable and poorly designed that even the large coils are worthless in the field. Professionals who make their living in the treasure hunting business always demand the best quality instruments and usually own at least two or more detectors capable of good field performance. After all, money missed is money wasted, and the deep ones you might miss would pay for many good instruments.

THE BFO

Without question, more actual treasure (buried and concealed caches) has been recovered with the BFO than all other detector types combined. I base this observation on my own thirty-odd years of experience and the fact that most early-day professionals used that old workhorse extensively. BFO's claimed the largest percentage of detector sales until recent years when coin hunting achieved its present popularity. The large number of weekend hobbyists made the standard TR with its lightweight 8-inch searchcoil and quick response the most popular for coin hunting. This is not to say that thousands of caches have not been recovered using detector types other than the BFO, only that it is obvious the BFO dominated the market in sales to professional treasure hunters. It is outperformed in many fields of use, but its ease of operation and versatility, plus the availability of large searchcoils, have made it a favorite of cache hunters for years. Regardless of more sensitive instruments that will go deeper and operate under certain adverse conditions with **fantastic results, the BFO can ALWAYS be safely used** or carried as a backup unit in most circumstances involving the recovery of buried caches. Its usefulness in cache/treasure hunting is attributable to its ease of operation around old homesteads; in weeds, grass and trashy areas; for building searches; in swamps and marshy areas; along rocky hillsides; inside caves and mines . . . you name it, the BFO has been there.

Your BFO choice MUST be absolutely 100% stable to achieve success, especially when using large searchcoils. The coils must be Faraday-shielded to prevent grass and weed interference. The BFO detector should be sensitive, but NOT at the expense of losing stability. Some units

This fellow seems to be quite happy about the early Spanish coins recovered on a South Sea island. You will note the old workhorse detector of the industry, the BFO. The BFO has probably been responsible for chalking up more treasure finds than all other detectors combined. Photo was taken in the late 1960's.

have the gain advanced so high to impress the customers with air tests that the units are unstable in the field and it is impossible to set the tuning speed at a moderate or slow motorboating sound. The slower tuning speed is necessary as you can hear the increase in beats better where you must move slowly because of limited maneuverability with larger searchcoils.

The medium size 12-inch searchcoil will operate best at approximately four inches above the ground, an operating height which frees the coil from most ground effects and obstructions. Tune at a slow or moderate motorboating sound (twenty to thirty beats per second); scan at the rate of approximately three feet per second. You may decide to change the scanning speed after becoming familiar with the faint increase in beats when over a metallic target. The 12-inch coil will detect targets as small as a fifty-cent-piece. It will have a maximum depth of approximately four to six feet on mass targets (in mineral-free soil). It is the best size to use in building searches; however, in small spaces it is sometimes necessary to use smaller coil sizes. The 12-inch coil would respond to a roll of coins concealed behind a wall. It is the best size to use for suspected caches similar in size to a small tobacco can.

A 12-inch coil does have limited use, however, because larger sizes will detect targets as small as a silver dollar *and* have the advantage of going much deeper on larger caches. Twelve-by-twenty-four, eighteen, twenty, twenty-two-inch, *etc.* . . . these large coils are certain producers. They are small enough to maneuver, in a practical sense, yet large enough to get all depth the BFO is capable of producing. Of course, you may use still larger coils but their maneuverability is limited. If the cache is thought to be deeper, beyond the limits of the BFO, then you should employ a VLF type with large searchcoil, using a BFO for a backup unit.

BFO searchcoils can be designed in many different configurations, round or oblong. The extremely large coils designed in round patterns present a problem as they are difficult to sweep widely enough to allow the deep target to respond. Elongated coils, such as 12-by-24-inch, weigh less and produce better target signals because the coil is narrow but still long enough for fast ground coverage. This size and type of BFO coil is best for deeper caches and it is produced by most BFO manufacturers. That is not to say the elongated coil will go deeper or even as deeply as the larger, round coils, only that it is more practical for field use and that the slight loss in depth is more than made up for by the sharper response.

We now have chosen the most practical combination available in the BFO line, a stable unit with approximately 12-by-24-inch searchcoil. This coil should be operated four to six inches above the ground, regardless of whether the soil is mineralized, to help free the coil from ground effects and clear most obstructions. The question may be asked, "Why not try to operate closer to the ground and gain more depth?" When the BFO is operating at a smooth, even beat, the deep or faint signals are more easily heard at the suggested operating height. If you operate the coil TOO closely to the ground, you will have erratic or uneven audio response which will more than offset the advantage gained in four or six inches. Anyone can become very proficient with a BFO if he takes the time to operate the detector where the beats are very *smooth* and *even*. The response to the slightest presence of metallic targets will then be very distinct.

Let's attempt to recover a suspected treasure cache from a typical location such as an old homestead, ranch house, saloon, ghost town, *etc.*, or in any area where there is probably an abundance of small metallic junk. Someone will likely say, "I can call my shots and tell the cache from the junk." Perhaps this is possible in some circumstances, but not always. Small junk could have

L. L. "Abe" Lincoln points to a spot by this old house where a treasure cache was once buried. For some reason its owner did not return to retrieve it. "Abe" is searching with a TR equipped with one of the large egg-shaped searchcoils. These coils are finding increased use by the TR operator, especially in areas that do not contain much ground mineralization. The TR coin hunter can accomplish a fair amount of cache hunting with these extra-large coils.

been scattered on top of the buried cache by accident. Ten to one the cache will be in a can, kettle, jar, or some other junk or ferrous container that would respond the same as all the other unwanted trash. Sometimes this non-profitable trash will have to be removed patiently and slowly if you are determined to make a recovery. The methods used and work required will vary somewhat from location to location. The 12-by-24-inch BFO coil will not respond to trash items smaller than a silver dollar. This characteristic saves much work and time. The BFO type detector is not super-sensitive to small ferrous iron so this presents no problem. You will be able to determine the size of the indicated target to a reasonably certain degree, depending on your experience. The twelve-by-twenty-four-inch coil will quickly cover an area and save time. If you wished, you could dig only the indications that you considered to be deep. These signals generally are very faint but, depending on your experience and confidence, you may be able to define them from surface cans and trash. If you were to attempt to search junky locations with detectors using the 6 or 8-inch searchcoils so popular for single coins, you probably would find many small relics and junk items. However, if you wish to cache hunt for treasure you should employ *large* searchcoils to obtain all the depth possible.

THE VLF

Perhaps research has convinced you a cache is actually in place with little chance of its having been moved. Maybe the location has much mineral content, or the cache could possibly be one of the very deeply buried ones. Under excessively mineralized conditions you have only one option: use a VLF type detector with large searchcoil. The VLF will penetrate the mineral; the large searchcoil will free the super-sensitive instrument from response to some of the extremely small trash items, such as small nails, pieces of wire, tin, *etc.* The presence of trash items is one reason you did not attempt to search the location with a VLF detector to begin with. The VLF type is extremely difficult to use in littered areas. If you are convinced the cache is there and the BFO could not locate it due to some terrain condition, then there is NO choice . . . the VLF *must* be brought in.

The VLF types are magnetic phase detectors with true mineral-free operation. The VLF, called by many different trade names depending on the manufacturer, is truly a deepseeker, not affected by adverse mineralized conditions. When a suspected treasure cache is located anywhere under reasonably isolated conditions, the VLF type will go deeper, operate perfectly and outperform any other detector type, *provided* you employ the large searchcoils available for these deepseekers. Do not become overconfident and rely on the medium size coils. If a cache is worth recovering it is worth the small added cost of a large coil to ensure that, if the cache is within detection range of ANY detector, you have the best and most efficient instrument at your disposal.

NOTE: The RF (two-box) type detector would pass small trash items and permit quick ground coverage, but it would not penetrate the mineralized background. Under these conditions the two-box RF model has been known to miss many cans, jars and other small caches because the mineral content forces you to detune ANY type of standard transmitter-receiver to permit operation. This will include attempts to use standard TR (IB) detectors with large coils attached. In highly mineralized areas they must be detuned to the point where they lose *considerable* sensitivity.

The VLF types should be used with the largest coil available; the *larger* the searchcoil, the less you will have to dig small nails and tiny pieces of ferrous iron. The larger coils on the VLF's are quite heavy, but they are better because they produce more depth and free the super-sensitive instrument from responding to some of the small, worthless and time-consuming junk.

Here is Dorian Cook constructing a metal detector proving ground where Garrett instruments will be proved during their initial design stages. The test plot is divided into two sections: mineralized and non-mineralized. In both sections targets are buried from six inches to six feet. Of course, testing and proving of a more serious nature and all final testing is accomplished in areas throughout the world where detectors will actually be used by treasure hunters.

Without question, all of us have left many deep caches that were beyond the detection range of the detectors we employed. Most professionals use the best detectors available and employ the large coil attachments. Even so, many caches, at a shallow depth but concealed under rocks, are missed. This can almost always be attributed to the detector's failure to penetrate mineralized rock. The loot or relics are still there, awaiting the next hunter who is using a mineral-free-operation type of detector.

Many failures can also be attributed to the hunter who may be thoroughly experienced as a coin hunter but inexperienced as a cache hunter. Because he has full confidence in his detector for coin hunting he overestimates its penetration on deeper objects. An erroneous interpretation may be arrived at by his having recovered a tin can or larger object at great depths while coin hunting. No consideration was given the terrain which was easily accessible with his coin hunting detector and small searchcoil, and no allowance was made as to the presence of a small amount of mineral content.

He then comes across information leading to a ghost

This collection of buffalo nickels represents the effort put forth by one treasure hunter. He found some of these coins while coin hunting; others he found concealed in a jar in an old building.

town or suspected treasure site, perhaps an old Spanish mission, site of a stage coach robbery, miner's stash, Indian cache, *etc.* He may spend considerable time doing all research possible, enlist the aid of a partner, take all kinds of camping equipment and spend a few hundred dollars for a weekend expedition. After this expenditure of time and money, he generally neglects the most important tool necessary for recovery of the cache— the correct *type* of detector, one capable of utilizing *large* searchcoils. If the area is mineralized (and almost always it is) a detector type that will *penetrate* mineralized rocks or ground is required. Perhaps his prized detector is a standard TR, so popular for coin hunting; perhaps it is a discriminating type capable of utilizing only a small searchcoil. No matter . . . he is almost helpless on deeper objects. In many magazines you will see pictures of groups searching in isolated areas that probably required many days of preparation and much expense to reach. You will notice oftentimes that the searchers are using small 8-inch or medium size searchcoils. These treasure hunts generally produce a few old coins or ancient relics but seldom a cache, especially one buried in rocky mineralized areas.

For your own edification obtain a GALLON bucket or can and a large slab of highly mineralized rock approximately four to eight inches thick and two or three feet wide. Dig a very SHALLOW hole and just barely cover the gallon can. (It does not matter whether the can has money in it . . . a metal detector only "sees" the outside metal object.) Slide the large rock slab over the buried can; tune your detector to the ground mineral content (not over the rock); and pretend you do not know the cache is under the rock. Start your search and pass over the large cache. You probably did not get ANY indication; if you did it was very slight. Consider that a gallon can will hold approximately $27,000 in gold at the old price and approximately $1,200 in silver dollars. (Understand this large a cache is not common as many consist of only a few thousand dollars at most.) Now we are talking about a SMALL can (for example, a tobacco can) or quart container, perhaps an old fruit jar with zinc lid or maybe saddlebags with a few handfuls of gold coins in them. Consider what will happen when the suspected cache consists of only a few thousand dollars in gold and may be buried anywhere from one foot to three feet deep. Large iron chests and kettles are not really TOO much larger than the gallon can and, if they are concealed under mineralized rocks or buried at arm's length in mineralized ground, you may now understand WHY you HAVE NOT been finding them.

Professional cache hunters understand all this and make allowances for the condition of the search area and the fact the cache may be deeper or smaller than antici-

A case in point on successful cache hunting . . . Bill Mason of Redwing, Minnesota, used a *large* searchcoil to recover this small but valuable cache. It had been passed over many times by coin hunters using smaller, coin hunting searchcoils. The smaller TR and BFO coin hunting coils simply do not penetrate deeply enough. Experienced cache hunters know this and use the largest searchcoil they can obtain to achieve all possible depth. Bill said he had also been guilty of coin hunting over this very same place and had received *no* indication from his 8-inch TR coil. The cache was simply too small and too deep for the smaller coils. Bill now pays careful attention to his choice of equipment for each situation.

pated. They take all the precautions they can, such as attention to *deep-seeking* instruments using the *large* searchcoils. The simulated cache under the slab rock and in mineralized ground would also have been very difficult (if not impossible) to detect with the BFO type equipped with large searchcoils. The VLF types provide your best chance for recovery in rocky or rough terrain where mineralization is present. However, in rocky canyons and slide areas that contain huge bluffs and cave possibilities, you must quickly learn how to operate your VLF type in order to make correct identification on the out of place magnetic iron indications you will receive from the VLF type detectors. These "out of place hot rocks" can be identified by correct detector operation and understanding what the various detector signals are telling you.

One of the major reasons you see very few cache hunters using the deep seeker types with small searchcoils is that the excitement of finding small coins does not appeal to them. The coin hunter will almost always have fun finding a few coins and related small objects. The cache hunter, however, does not desire these small objects and may sometimes go for months or perhaps years before making a recovery, a situation which can become frustrating. Of course, the experienced hunter KNOWS that if he really persists, ignoring the taunts of his fellow hunters, he will sooner or later score. This does make a difference as the cache will probably more than compensate him for his diligence.

I am personally acquainted with two nationally known hunters who, when they first started treasure hunting, carried only a TR detector with 8-inch searchcoil. Through advertising and other means they gained many good treasure leads and information relating to suspected cache sites. They expended many thousands of dollars and much travel time to investigate these leads. They did recover many coins and shallow relics, but think of the caches that were passed over and missed due to the confidence placed in the small searchcoils! I know many experienced cache hunters who would have paid good money for some of this information and leads or who would have worked them on a percentage basis.

Never pass a suspected treasure site just because someone tells you it has been worked before. More treasures are missed than recovered, all due to the depth of burial and mineralization content in the ground. Consider old parks where coins are found; the number of years they are hunted by thousands of coin hunters; and the fact they never become completely hunted out. The same is true of suspected caches. When better detectors come along, these old sites should always be promptly reinvestigated. If you do not, someone else will. Many recent recoveries have already been made from old locations since the introduction of the mineral-free operation, deep-seeking VLF type detectors.

If you are searching in non-mineralized zones, such as a few areas near the coast, and you already own a standard TR (IB) type detector and consider the cost of a deep-seeking type unwarranted, you may achieve excellent depth by using the extra-large searchcoils available for the standard TR. Try to obtain coils larger than the standard 12-inch that comes as an accessory on some models. These extra-large TR coils are quite heavy, but they will penetrate the mineral-free soil and you will probably have excellent stability of operation. Of course, you will still NOT achieve the depth as with the VLF or PRG types, but the necessity of another detector for specific mineral-free areas will be eliminated. Attempts to search in highly mineralized zones will meet with instant failure due to erratic detector operation, but you will at least be able to make your coin hunter TR double as a deepseeker under favorable conditions. The extra-large size of the searchcoil allows you to by-pass many small troublesome signals received from coins and small junk while engaged in the search for deep caches, large cannonballs, battlefield relics *(not* small bullets), old bottle dumps, and the like. Of course, ANY type detector that permits attachment of large searchcoils will perform admirably under such favorable circumstances. The VLF, PRG, and RF models will all produce approximately the same depth in mineral-free ground. The VLF will produce the *most* depth in mineralized ground, but, if you own a good TR capable of using extra-large searchcoils, *in mineral-free ground* you can also achieve excellent results.

THE PRG

The PRG type detector will permit the attachment of large searchcoils. As mentioned previously, it has almost perfect target identification and depth equaled only by the VLF and RF (two-box) type in non-mineralized zones. It does take two or more operators to use the large searchcoil that is available. It is impossible to be precise, but if the suspected cache site were in a salt water area the PRG equipped with the large coil would probably detect more deeply than any other type detector and produce almost perfect target (the container in which the cache is buried) identification. Of course, its use is limited to the search of areas relatively free of iron mineralization and it is quite costly, especially if the large searchcoil is included. However, cost is never a factor with professionals who consider only the possibility of success.

You should not consider all salt water areas to be free of iron mineralization. Most ocean beaches are saturated with black sand (mineral) and are highly negative. The tide comes in, wets the mineral (black sand) with salt water, and the area becomes conductive (positive) to all detector types except the PRG and PI. The tide goes out; the beach sand dries. It then becomes highly negative (mineralized), and the PRG becomes unproductive. In low-lying marshy, tidal zone areas the sand never dries out. Here the PRG is rated tops in discrimination or target identification, equaled in depth only by the VLF and perhaps the PI (pulse induction) types, provided you use the same size searchcoil.

It would be well to dispel confusion as to mineral salts that become conductive (metallic response) and mineral (Fe_3O_4, magnetic black sand) that produces a negative response. Both are considered "mineral," but the salts become conductive *only* when wet. The VLF type can be adjusted to operate mineral-free over any amount of negative black sand (mineral, Fe_3O_4). The ground or terrain control will simply zero out the magnetic black sand and permit full depth penetration. The PRG cannot be adjusted to operate on the black sand, but it has the advantage of target identification in salt water areas.

If your suspected cache location falls under any of these specific situations and cost is not a factor, consider the PRG type to be the best "discriminator" in detector instrumentation, especially with regard to man-made ferrous iron.

THE PI

The PI type of detector has been discussed in preceding chapters as to use and sensitivity. This supersensitive instrument will also penetrate locations that contain a high degree of mineralization. It has the disadvantage of being overly sensitive to small ferrous iron (the same as the VLF type) and most models do not permit the application of large searchcoils. It is much harder to use for pinpointing as the "after-ring" of the audio signal is very confusing in littered areas. It costs considerably more than the VLF types (provided you purchase models with the large searchcoils) and does not produce any added depth to offset these expenditures. This type detector is excellent in many circumstances, but it is far too troublesome and time-consuming to use around suspected cache sites in trashy or littered areas.

The recovery of deep caches requires the use of large searchcoils. All models of this type do not accommodate this added accessory, and care must be taken to choose the model with large coil capability. In mineralized areas it is comparable in operation sensitivity to the VLF type. It is super-sensitive to ferrous iron. When the large searchcoil is used, it is very hard to narrow the target area and pinpoint accurately. Models permitting the application of large coils are *much more costly* than the VLF types and, considering that the PI has no more depth or versatility, it is hard to justify the added expense. It is imported from England and distributed throughout the United States. It is a quality instrument, but does not perform quite as favorably in applications such as cache hunting as do American-made detectors. If you desire a PI for cache hunting, I suggest you contact a U. S. distributor for more information. They always give true and explicit information in regard to this product. I have used these PI instruments and under certain circumstances they are very effective. However, as in the case of the VLF type, they MUST be backed up with a BFO detector in highly mineralized zones. They produce the same false signals on extra hot or highly mineralized rocks that are out of place with regard to the surrounding matrix. The oft-quoted remark, "Specialized instruments should be used only where they excel" seems to hold true, especially here.

THE RF

The RF (two-box) is one of the oldest type detectors in production. It definitely has a field wherein it excels: *pipe and cable location.* It does not need large search-coils, for its construction allows the utmost in depth penetration. It is adversely affected by magnetic iron (mineral content of search area). It produces more false signals than the VLF type as it responds to many of the earth's anomaly changes. I personally have carried this type many hundreds of miles, on many different locations, and used it under many different soil conditions. It has the advantage of responding ONLY to targets that can be considered larger than a baseball. This characteristic permits the exclusion of response on most small trash items when cache hunting. If the soil is relatively free of mineral content the RF will produce great depth and fast ground coverage, an added advantage when cache hunting anywhere.

The only drawbacks are if the soil contains excessive mineral or if the location is in a rocky area. In such cases the RF must be detuned to the point where it loses its depth penetration and becomes almost useless. The two-box also responds to both metal and mineral changes, leaving the operator at a complete loss as to correct target identification. This problem can be partly eliminated by application of a BFO type to identify some of the more shallow indications correctly. It is highly productive in flat terrain where mineral is not abundant. Over the years many large caches have been found with this type. Do not attempt to use the two-box RF in salt water zones or marshy areas as the depth is *drastically* reduced.

One of the older, more successful treasure hunters used one of these two-box detectors with great success, and I have made my share of recoveries with one. However, since the recent production of the VLF types by many manufacturers, the two-box RF is no longer necessary as a deep-seeker. VLF types will actually detect smaller targets in mineralized rocks and soil, and they have the advantage of complete versatility in operation. The application of extra-large searchcoils on VLF types produces as much or more depth in mineral-free ground and *much* greater depth under highly mineralized conditions. I and many other cache hunters will hate to see the old two-box deepseeker retired as a cache hunter, but electronic facts cannot be changed. I will miss digging some of those deep, empty holes that were so common. If you own one of the RF models perhaps these field tips will assist you or, if you intend to purchase one, use it only where it excels . . . locating buried pipes and cables.

SUMMING UP

To sum up, areas that have had considerable habitation and are suspected of harboring a deeply buried cache should best be left to investigation with BFO detectors with large searchcoils. Use of the VLF type employing a large coil may be practical or necessary under certain circumstances, as may be an occasional use of the RF (two-box) in some rural or farm areas where mineralization is absent. Yet, even in these circumstances, the BFO must be used as a backup detector as the two-box will respond to underground seeps and earth abnormalities and create many false indications.

I, as well as many others, have spent years searching for suspected caches on rocky hillsides, canyons, rock bluffs, talus slides and river banks with gravel concentrations — areas far removed from recent habitation and that probably contain only small amounts of trash, though they may be highly mineralized. Mostly we have relied on the BFO and RF (two-box transmitter-receiver) types; however, a few successful hunters have employed the early-day horizontal loop TR detectors in certain areas. The fact that huge rocks and canyons are mentioned does not mean all rock contains the magnetic iron that is so disastrous to effective TR operation. A few isolated areas in the bluff or canyon section of Utah, Colorado, Arizona, New Mexico and elsewhere are of limestone or sandstone composition and may contain no magnetic oxides. As mentioned previously, probably ninety percent of the entire earth's surface contains iron oxides to some degree. The earth is thought to be solid, consisting of many minerals and metals, but only magnetic iron oxides create havoc for the metal/mineral detector.

Many failures can in part be blamed on the section of the country in which the treasure hunter lives and also on the reluctance or inability of a detector dealer to explain that a model he sells does not have deep detection capability or the versatility required for cache hunting. In certain states some particular type of detector may have achieved widespread popularity. This may be because local dealers stock this specific model or brand almost exclusively. Perhaps there is lack of public awareness that successful cache hunters operate in the area. A few dealers are only sales motivated (or perhaps just plain inexperienced) and do not bother to explain that another type of instrument would be better suited for your particular job requirement. This combination of conditions sometimes leaves the cache hunter who wishes to become successful no place to turn, except perhaps to knowledgeable books on the subject. Even books are not the final answer. They may be written to promote a particular brand or type of detector, or they may be written by an armchair treasure hunter who has had limited field experience. All types of detectors should be explained and presented in an unbiased manner. Field tests are always the buyer's best protection and guide. Like the man once said, "By the time you gain experience, sometimes you are too old for it to do you any good."

USE A PROBE

Always carry a time-saving steel probe to use where the soil permits. If you think the response from your detector indicates a target large enough and deep enough to fit your idea of the suspected cache, probe the spot before digging. Experienced operators can probe carefully and determine what the target is. A glass jar would be easy to define. The depth at which it was buried would be an indication as to when it was placed there. If the probe hit a flat piece of tin and you forced the probe on through and felt nothing else, the target would probably be just that . . . a flat piece of junk tin. If you hit a tin can and the probe penetrated so that you could tell there was something in the can, you would want to dig it. Many cache hunters who use probes become so proficient with them that they can feel a newspaper when the probe passes through it.

Of course, a special kind of probe rod is required. A picture and the instructions for making one are included herein. The probe has a steel ball-bearing welded onto the point end to permit the probe to move up and down with no restrictions so you can define more clearly what the underground object is. If the ground is rocky or too hard a probe is not helpful. In that case you will just have to dig all indicated targets that you judge deep and large enough to be the suspected treasure cache. (You may wish to use a backpack when in the field, as I do, either enroute to a location or coming from one. It helps avoid drawing attention from the curious because I appear simply as a hiker or I am just enjoying the scenery . . . plus the pack is a convenient way to carry equipment and supplies. Large searchcoils will generally fit into the large backpacks, as well as small shovels, detector, probes and necessary tools to make an average recovery.)

THE TRUE CACHE HUNTER

There are coin hunters, ghost-towners, relic hunters, and then there is the cache hunter. He or she is generally not interested in a few relic items and a couple of coins which probably would not even pay for the gas. The cache hunter is interested in finding loot — a bundle of it! Of course, the cache hunter does not find such a target every day. The beginner must realize this and not become discouraged. However, there are millions of dollars stashed in the ground and, if you persist, sooner or later you will hit a cache. A lot of people have become wealthy from pursuit of this interesting occupation.

The TRUE CACHE HUNTER is definitely a separate breed in the treasure hunting fraternity. This person is seldom seen among weekend hobbyists, those who hunt for coins and relics just for the fun of it. I am personally acquainted with two highly successful cache hunters who NEVER coin hunt, preferring to spend all their time in pursuit of larger, more profitable finds. I am sure neither

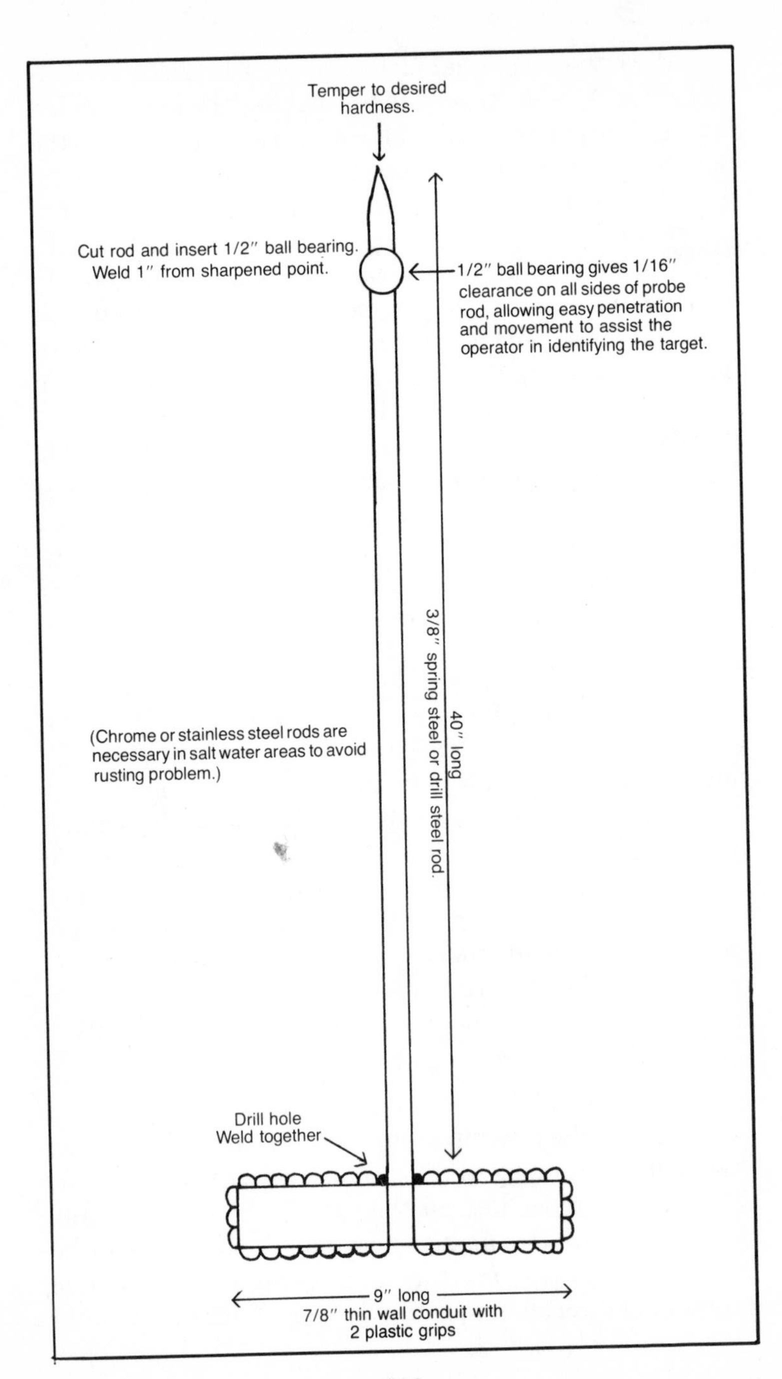
Temper to desired hardness.
Cut rod and insert 1/2″ ball bearing. Weld 1″ from sharpened point.
1/2″ ball bearing gives 1/16″ clearance on all sides of probe rod, allowing easy penetration and movement to assist the operator in identifying the target.
40″ long
3/8″ spring steel or drill steel rod.
(Chrome or stainless steel rods are necessary in salt water areas to avoid rusting problem.)
Drill hole
Weld together
9″ long
7/8″ thin wall conduit with 2 plastic grips

considers coin hunting as a hobby to be beneath his dignity. It is just that small object searching never appealed to them. (This is true of the big game fisherman who spends his entire life in the search for record-breaking marlin or sailfish.) Most, if not all, real cache hunters spend much of their time in research, seldom mentioning their occupation to anyone other than perhaps another professional. Their actual expenses are considerable as research sometimes demands extensive travel. They may occasionally have to pay a rather large sum of money to obtain information. Perhaps they may even offer to share the cache on a percentage basis, a very common practice, especially for permission to search on private property. Research is very time-consuming, and it is advisable to work on more than one project at a time if possible. Due to differences in soil conditions it is many times necessary to purchase a special type of detector for a particular location. Patience, as well as financing, is required.

It is best to maintain a low profile when engaging in research or in an actual recovery attempt in order not to be bothered by the curious or those not actually entitled to a share of a recovery. NEVER, NEVER put your trust into a *verbal* agreement with a landowner and NEVER leave an open hole after you have removed something. A case in point . . . one of my very good hunting buddies who has made more than his share of actual cache recoveries left a hole open about twenty-five years ago. He had recovered a Wells Fargo box from the hole. The open hole left an impression that something had been removed from it. The landowner had given permission for entry and search but, upon observing the empty hole, suddenly changed his mind. Thoughts of vast wealth probably flooded his mind, and he immediately contacted a lawyer. The lawyer advised a quick civil suit for recovery and damages. You understand my friend's predicament. He transferred his real property to a relative's name and let the lawyer fume as there was nothing on which to file an attachment in court. The landowner did not want to pay for a lost cause so after a few years the incident was forgotten. My friend, the professional cache hunter, learned a valuable lesson he never forgot.

The same holds true of arrangements with partners. It is only fair always to have an agreement in writing in proper legal form. I learned this in 1947. Two other hunters and I were on to an old sheepherder's cache that we finally recovered. One of the partners was an inexperienced local gentleman, taken in on an equal basis, to help find the suspected location. The find was close to the five-figure mark, and when he saw it he became unreasonable. He demanded and got half of the cache; we had no choice. This convinced me to "get it in writing" and to deal only with people who had already seen "found" money and recovered it, people who would be able to conduct themselves in a businesslike, professional way. Many of my friends who are professionals in the business have had the same experience. Generally, you can trust a cache hunter who makes his living in the business. For one thing, he CANNOT AFFORD to have his name besmirched in his trade and, for another, he has handled found money and does not get excited and perhaps think of trying to take your share.

As for obtaining permission to enter private property, *of course* you should seek permission. If a cache was buried long ago there may be a problem in determining the legal ownership of the property where it is buried. WHO does it belong to? The land belongs to an individual, the state or the federal government. According to most state laws governing treasure trove, it belongs to the finder; however, you may be guilty of trespassing if you do not secure permission to search. The cache does not become treasure trove until it is found. If the find must be turned over to the authorities to decide ownership, the case may be tied up in court for a long period. When all is said and done, lawyer and court costs may take up the biggest share.

If the land is posted, use sound judgment as to whether you wish to wind up in court or whether you should first obtain permission to search. Cache hunting can be illegal; so can driving down the road. It would be difficult to go forth in the wide open spaces and fail to break, knowingly or unknowingly, some particular law. I believe in respecting all property rights. If you treat property owners with courtesy and friendliness (if you are trespassing) they will more than likely not evict you. There is no hard or fast rule. Many states have a permanent trespass law which can be very strict. At my age, I consider myself too old to have to lie, too smart to steal, and too fat to work . . . so I will just continue cache hunting and allow myself the privilege of telling it like it is.

TAXES

Pay what is due. Do you think you can find $20,000, buy a new car, take a long-awaited vacation, etc., WITHOUT the extra income being noticed? Hardly! So declare the extra cache money as unexplained income, pay your fair share and rest easy. Of course, you owe nothing until you convert the value of the cache into spendable money; then it becomes legal, taxable income. If you are making money in the treasure business it will not break you to pay properly due taxes. If you are among the many professionals who make their entire living at this you will be able to deduct expenses incurred while engaging in research and recovery. It is advisable to list your income properly. Many states have explicit laws governing treasure trove and can claim part or all of everything. If you declare extra income, whether from gambling, horseracing, found money, *etc.*, it should not be anyone's business as long as it is not stolen. However, many states are quickly passing laws to recover all the income they can from treasure hunters who have invested time and money (sometimes their lives) in reclaiming the lost from oceans to the highest mountaintop. Don't draw unnecessary attention to yourself, pay your legitimate taxes, and insist on your rights.

A large national organization that is dedicated to fighting for the rights of the outdoor individual, both the treasure hunter and the recreational miner, is GOLD PROSPECTORS ASSOCIATION OF AMERICA (GPAA), P. O. Box 507, Bonsall, California 92003. George Massie, President, will gladly send information on how to join and obtain a free copy of their nationally distributed publication. The Association has the membership strength to work for the individual and maintains a question-and-answer service for all GPAA members. There *is* strength in membership. Many of us in the business are members.

For those of you who would like to delve further into the fields of research, recovery, tax problems, equipment, advice, and gain a thorough insight into the pro's and con's of professional treasure hunting, purchase a copy of *TREASURE HUNTER'S MANUALS #6* and #7 by Karl von Mueller, published by Ram Publishing Company. They are two of the most complete and most informative books on these subjects.

CHAPTER 13

General Treasure Hunting

Treasure hunting and the treasure hunter — what a vocation and what a person! Treasure hunting covers such a large scope of activities and such a large portion of the hobby field that it is impossible to define the phrase properly. Actually, the "old pro" treasure hunter of yesterday is fading, being replaced rapidly by the everyday hobbyist. There is no end in sight for this hobby, perhaps one of the most fascinating and interesting pastimes ever to capture the imagination of the world. More and more sportsmen have purchased a metal/mineral detector to add to their sports gear.

Consider the fisherman who purchases a metal detector to carry with him on his fishing trips. He is not a treasure hunter in the strict definition of the word. He is a hobbyist interested in outdoor activities. Most every sportsman who enjoys the great outdoors will eventually try his hand at this intriguing hobby. It will fill gaps left in his regular sports activities and provide added enjoyment for all members of his family.

Treasure hunting can be coin hunting, prospecting, bottle hunting, ghost-towning, and many other efforts to locate metallic items by electronic devices. Legislation and new laws to curtail the activities of treasure hunters have caused the hobby to emerge in a new light. You are not a treasure hunter when you attempt to find old coins; you are entering the fascinating field of coin collecting and coin hunting as a hobby. This will in no way deter or hinder you in the pursuit of other hobbies and sports activities.

Quite often Mr. L. L. "Abe" Lincoln comes through Lewiston, Idaho, to see me as he is on his way into the remote areas of Idaho and southern Canada to work early-day gold mining camps. On one such trip he brought several pictures showing some things he had located in New Mexico. This religious artifact, obviously of crude workmanship, was among the things he found. By dating some of the other artifacts found in the same box, it was determined that the cross was probably made in the 1600's. I thank "Abe" for letting me use the picture.

PROSPECTING/GHOST-TOWNING/COIN HUNTING. There is no end in sight for the hobby of treasure hunting! It is perhaps one of the most fascinating and interesting pastimes ever to capture the imagination of the world.

CHAPTER 14

Detector Operating Tips

CLAY DEPOSITS

These troublesome spots can play havoc with a metal detector. You will be using the larger searchcoils, covering a field or flat area, perhaps even a hillside. Suddenly your detector will give a metallic indication. It will generally be rather faint, like that for a deep cache. After digging a few feet you will discover an "upshoot" or chimney, a deposit of beautiful clay. If you pay close attention to the condition of the clay you can tell whether it has been dug before. If it has been dug, it will have streaks of black dirt or topsoil mixed in it. It would not matter how long ago — two-hundred years or more; it would still show foreign dirt mixed in the clay. If you check some of the clay with your detector you may receive NO response and perhaps consider digging more. If the clay shows it has not been dug before, dig no further.

This type of clay deposit is known as *neutral* clay. It has no content of mineral (black sand) and nor of metallics. The metallic response was caused by a vacancy reading because the detector was tuned to the surrounding terrain which had a mineral, negative response on your detector. When you passed over the clay deposit, the detector simply had a chance to gain or faintly increase in volume. The same response would have occurred if you had gone over a tunnel or cave, provided it had been within detecting distance and in a mineralized area.

A *positive* clay deposit may contain some type of natural conductive metallic substance and/or mineral salts that have become conductive due to the presence of moisture. The conductive nature of the wetted salts can cause the detector to respond faintly, as would a deep cache. When you remove a few shovels of this material and test it with your detector, it will respond positive, as would metal. The response will not be so strong as for pure metal (like a tin can or piece of larger metal), but strongly enough to positively identify the clay as conductive. This is not a mineral reading as most operators think but a true response on conductive substance. Check the clay carefully. If it has not been dug before and has no streaks of top soil mixed in, do not waste your time digging. The only time this response could be in error is when you are in a mineral or mining belt.

In certain areas it is possible that you could be receiving a response from high grade gold ore or other precious metals down below or mixed in with the clay deposit. In some specific areas gold deposits may be formed by a leaching process over many millions of years and rich deposits may be found where it came to rest against a retaining wall, forming a pocket. This crystalline type gold has been discovered in mud pits and many other formations of clays, minerals, *etc.* The famed crystalline wire gold of the Cascade Mountains in Canada, Washington and Oregon is a case in point. Prospectors who have not searched this area previously can hardly believe it when crystalline wire nuggets are found in muck and mud, high up on the mountain sides. If the clay shows TOO MUCH conductive content, it might be wise to have it assayed. It could possibly contain some rare earth minerals and have high market value.

TREE ROOTS, STUMPS, GREEN LOGS

Such objects may cause a situation where no hard and fast procedure can be advised. Many operators pass their detector over a tree root, get a slight positive signal, and quickly decide the detector responded to the root. In some cases this is true, but not always. I will explain in detail what can happen and the circumstances which may cause it to happen.

If you are conducting your search in mineralized (negative) ground and pass over an old dried root or place where one has decayed, you will get a slight metallic response, NOT from the root but from the vacancy created by the root. There was an absence of mineral in this spot which allowed the detector to gain in tone or beats with the same effect a metallic target would have produced. You will get such a metallic response from spots where old fence posts have rotted and left a vacant hole.

You might also get this same metallic response over a *green* root. This kind of response is always relative to the content of the mineral in the soil. Many roots do not respond as metallic because they are not dense enough or do not have enough permeability effect to interrupt the electromagnetic field produced by the searchcoil. This is why green tree roots in certain areas with different mineralized soil content may respond either way. Perhaps you wish to test your detector on certain tree stumps or trees. You will notice that the part you are testing is completely above ground and there is no chance of a vacancy reading. You MAY receive a slight positive or metallic signal and, of course, you know the tree is not metal. The permeability of the dense material (combined with mineralized sap) was greater than air and the electromagnetic field was interrupted, causing the positive increase in tone. This response can occur with both BFO and TR types.

When searching around big trees, there may be a once-in-a-lifetime false signal when you detect a large root. This is a rare occurrence so do not become alarmed when it happens. Your detector is performing correctly. The signal is caused simply by soil conditions combined with the mineralized sap in the root. I first noticed this effect many years ago when I attempted to find nails and small metallic objects in logs. To save their saw blades many sawmill operators tried to use a standard detector to locate foreign objects in logs before milling. The metallic target could never be detected very deeply and sometimes was *undetectable* if it was just under the surface of the log. We called the effect "log-itis," and sawmill operators were forced to have a special type of detector built for pre-milling scanning.

Because of the positive indication received from certain green tree parts many electronics engineers will quickly say the searchcoil is unshielded. (You will get a slight positive response from your hand on unshielded searchcoils.) This is not the case. I have used fully shielded searchcoils and received this same erroneous response in many different areas of the U.S. A fully shielded searchcoil will protect against false responses from grass, weeds, fingers, *etc.* There is *one* thing, however, that will respond as positive on even the best shielded coil . . . large FERNS. Ferns grow in shaded, damp areas. A fern draws the soil minerals into itself (it is hollow) and, when you touch the shielded coil against the fern, a *positive* indication will be produced. Green grass and weeds touched to the same shielded coil will produce little or no response. If they do produce any response it will be *negative*.

TR searchcoils can respond in two different ways on tree roots, depending on the size searchcoil you are using. Sometimes the transmitter portion happens to catch the root just right and see it as positive; sometimes the receiver portion sees it as negative. In the same way, the TR searchcoil may respond to mineral (Fe_3O_4) either way, depending on the amount of mineral, moisture content, depth, and the portion of the coil to which it comes closest. VLF and PI type detectors do not have the previously described response on green lumber. They penetrate the green log perfectly as they do not see the difference in permeability or intensity. Other than the special sawmill detectors (so-called "moving" detectors), VLF and PI types are about the only ones which can safely be employed to detect nails in green lumber.

When digging under a large tree and you encounter a

large root, do not assume too quickly that the signal you receive is false. A cache could have been buried prior to the growth of the tree and be BENEATH the large root. If you decide to dig, test down in the hole with your searchcoil to determine if the response becomes louder as you go deeper. A root rarely responds to this test. A signal from a root generally fades away after you begin digging. A metallic cache would cause a louder signal as the searchcoil approached it. Many years ago I recovered a cache under these same circumstances. I became disgusted when encountering a large root on my first attempt at recovery, decided a false signal had been produced, covered the hole and left. After thinking about it, I decided to go back, chop through the root and settle the question. I made the recovery and learned a valuable lesson.

ISOLATED SALT DEPOSITS

Small salt-saturated spots may occur inland where ocean salts are absent. In neutral or negative soil a small area that has accumulated excess salts over a long period of time may become conductive when wet or damp. This situation can occur anywhere in the U.S., but you will notice only faint metallic indications when cache hunting with the larger searchcoils, depending on the water content of the soil. The faint signal may cause you to dig, thinking a deep cache is there. There is almost no way to determine the correct situation until you check the condition of soil removed from the hole. If it shows streaks and signs of having been dug before, keep on digging. You may never encounter this odd situation, but if you do you will long remember it.

I have dug empty holes in many places. One was because of a false metallic response received while searching an old abandoned barn. A large cache was reputed to be buried in the immediate vicinity. After much research, two other cache hunters and I decided to search the old barn. We carried two different types of deepseekers, a large two-box RF and a BFO with a large searchcoil. We received a faint signal in one of the horse stalls, much as a deeply buried milk can might give. After checking it with BOTH detectors we decided we had found the cache. The earth was removed until we struck bedrock, about four or five feet down. We checked the soil that had been removed . . . no response We refilled the hole, checked the target area again, and received the same faint metallic response. We dug again, thinking perhaps we had missed it — nothing.

After drilling the hole for the second time I finally took note of our surroundings. Apparently the urine of the horse who had stood in the same stall for years had thoroughly saturated the soil with salts and the spot had become slightly conductive, the same as does a beach from ocean salt. The surrounding soil was mineralized (negative) and the detectors were tuned to the soil content. When we passed our detectors over the mass of damp conductive salts, the response was made much more positive because of the surrounding mineralized soil. I have come across this kind of situation only one other time in all my years of cache hunting. The location was in a cow pasture where a salt lick had been placed. This was years after the first occurrence so I readily understood the response, but I generously allowed my partner to dig to convince himself. There were some slight visible indications that the old salt lick had been in the immediate area or I would not have been so fortunate as to recognize the signal quickly for what it was . . . a false indication that certainly sounded like that from a deeply buried cache.

Perhaps the phrase "faint indication like a deep cache" is confusing. However, if you will question truly experienced cache hunters, they will be the first to say that all cache hunters continually listen and pray for that "faint" signal that means a deep target. A faint signal generally indicates the object was buried long ago and is not simply some recently discarded or lost item. Also, I

"Abe" Lincoln, Rogers, Arkansas, one of the most experienced cache hunters in the profession, is certainly one of the more successful ones. "Abe" never misses those faint indications. He knows they mean there is a much greater chance the target has been buried a long time and less chance that it is recent trash. "Abe," a dear friend, moves slowly and carefully, always thoroughly investigating even the slightest indication. That might explain why he is so successful!

think most searchers tend to miss the very deep targets. (I do not mean coins; I mean a deep cache.) This can be attributed mostly to the searchers not having enough experience with extra-deep targets and, unless they are in the business, many do not get to conduct enough actual cache hunting projects to understand all the different signals correctly.

MINERAL AND THE METAL DETECTOR

The misnomer "mineral" in relation to a "metal" detector has created so much confusion in the minds of treasure hunters and prospectors that I believe some explanation is necessary. Actually the so-called "mineral" tuning on any type of detector is not absolutely necessary. Let's give a verbal illustration and say that on your radio you have two different stations with just a small silent space between them. If you set the tuning in the "dead" or silent space between the stations and slowly turn the tuning knob (clockwise) to receive one station, we can call that one the *metal zone* (positive). If you slowly turn backward (counterclockwise) from the positive one and cross completely through the space (silent zone or null area) where you first started, continuing to turn and tune in on the other or opposite station this would be the *mineral* (negative) side. This is exactly what the mineral side of tuning is on a metal/mineral detector, THE NEGATIVE SIDE OF POSITIVE. When using your detector, you would achieve the same results by tuning in the metal zone with a slightly audible tone or beat emanating from your speaker. As you pass over mineral (negative) the tone or beat would decrease or stop. You are then alerted to the fact that you have passed over magnetic iron oxide (Fe_3O_4).

There is no absolute necessity to require mineral tuning (negative side) on the detector merely to alert you to the presence of non-conductive iron in its natural state. The entire earth's surface contains mineral (magnetic iron oxides) to some degree and causes a "negative" response to your "metal" detector, so for all practical purposes you are detecting only good old Mother Earth. Of course, in isolated circumstances you might wish to locate veins of magnetite containing a high content of magnetic iron oxides. This should be attempted only with a BFO detector as, because of its uniform searchcoil pattern, this type is the only true metal/mineral detector. The prospector relates the "mineral" to that for which he is searching; the treasure hunter regards it as something in which he is not interested. Nothing could be further from the truth. If the prospector attempts to use the *metal* detector for prospecting he will almost always be searching for metals, NOT negative mineral (non-conductive iron). Gold is a METAL, not a mineral, and all types of non-ferrous metals such as silver, copper, cinnabar, tin and man-made iron objects will respond as metals, provided they are in a conductive form. So the prospector or anyone else is actually using the *metal* detector to search for metals.

For all practical purposes, the TR type metal detector is just that — a *metal* detector. Regardless of whether the mineral side of the tuning is on or off, the word "mineral" does little more than confuse the operator. Many manufacturers have completely done away with the mineral side on their tuning control, thus enlarging the silent or null zone. This does make it slightly easier for the operator to tune the detector correctly. Actually, the TR type has two or more windings in its searchcoil, a transmitter portion and a receiver portion. Regardless of whether the windings lie in parallel (induction balance) or are stacked (co-axial), the coil cannot be used to identify ore specimens correctly, either conductive metallics or non-conductive mineral, Fe_3O_4. Because of the two or more different portions involved in coil construction a TR coil simply cannot produce a uniform pattern on a target in close proximity to its surface. The coil is affected by size of the target, moisture content, mineral content, width and length of the target in relation to the searchcoil, and other factors that *prevent accurate identification of metal vs. mineral*. Thus, it is a fallacy to continue to refer to any TR type as a "metal/mineral" detector. The TR type is truly the finest and most practical instrument available for the detection of small metallic objects, a real *metal* detector in every sense of the word, but NOT mineral.

The BFO type detector searchcoil does produce a uniform pattern on targets in close proximity to its surface. The BFO type is ALWAYS constructed with both metal and mineral modes of tuning. Again, however, there would be NO ABSOLUTE NECESSITY to indicate a "mineral" side of tuning. It could be considered helpful in the search for black sand pockets, dredging operations, and rockhound identification of some gems with high iron content. You would, however, *still* receive the same accurate identification when the detector is tuned in the "metal" mode. If the beats decrease, it simply means you have passed over "negative" mineral. If you use the detector to prospect you are still searching for *metals*. Any decrease in beats or tone means only that there is a predominance of magnetic iron oxides, NOT necessarily the precious minerals the prospector thought he would find with the detector.

If you are treasure hunting you are definitely searching for metals. The mineral side of the tuning has no bearing on establishing the BFO as an all-purpose detector. The wide dynamic operating range, the constant audio response, and the fact that the BFO searchcoil produces a totally uniform search pattern to permit correct metal *vs.* mineral identification are characteristics which make the old workhorse an all-purpose, indispensable instrument. You would still receive the correct identification of metal *vs.* mineral if the detector were tuned in the metal mode of operation and you passed over magnetic iron in its natural form. The tone or beats would simply slow down or stop. Whether or not the tuning has both

"Hardrock" Hendricks, one of the more experienced prospectors, employs electronic prospecting methods. "Hardrock" is using one of the mineral-free VLF type detectors in a dry wash. If there is a gold nugget in the old creek bottom anywhere within detectable range, "Hardrock" will find it. He is familiar with the VLF types and knows he may have to investigate a few false indications which could be caused by a small rock of high iron content having been deposited in the dry wash. The time spent investigating a few erroneous responses is a small price to pay if there is a large gold nugget present . . . "Hardrock" and the VLF will locate it, regardless of magnetic iron content almost certain to be there in abundance.

"metal" and "mineral" incorporated into the system, the BFO type is a TRUE METAL/MINERAL DETECTOR.

MINERAL AND THE VLF TYPES

I consider the lack of information and instruction regarding this type of detector sufficient to warrant a brief explanation of the tuning system. The VLF type has been thoroughly explained in preceding chapters as to limits in usage, advantages, adverse soil conditions, *etc.* There is, however, almost a total lack of information or instructions available to the operator as to WHY and HOW the tuning system works, what causes false signals, and how to prevent or at least recognize them for what they are. I consider the tuning as one of the most critical phases of operation of this type detector. If it is fully understood by the operators most disadvantages can be readily overcome. (As mentioned in the chapter on detector types, the designation VLF — Very Low Frequency TR's — includes ALL trade names such as GEB, VLF, MPG, MF, MFO, TGC, plus more model names sure to be forthcoming.)

In the previous paragraphs on "Mineral and the Metal Detector," the two zones of tuning, metal (positive) and mineral (negative), and the silent space or null area BETWEEN the two were discussed. For all practical purposes, the same principles are true of all detector types. They have a metal zone and mineral zone with the silent area between the two zones. Many TR manufacturers simply leave off the "mineral" zone designation. The silent area is included in this space and the entire silent area is referred to as the "null" zone. The mineral zone is still there electronically, but an audible response is not produced when the tuner is rotated in that direction. On standard TR's, IB's, RF's, PI's, or PRG's the null area remains in the same spot, and the threshold of entry into audible sound remains the same. On the VLF types there are TWO tuning adjustments. The control called the terrain (ground or zero) control causes the null area and the threshold point to change or shift when the instrument is balanced or tuned to eliminate the effects of either negative or positive mineral. The other control is simply a tuning device common on all detectors which is used to increase or decrease the amount of audible sound.

Obtain a standard twelve-inch ruler and place it on a flat surface facing you, reading inches left to right. Posi-

Tommy T. Long (right), Boise, Idaho, is using a VLF type detector to locate a small conductive ore stringer. Allan Cannon, Pomeroy, Washington, is carrying a BFO type, complete with large searchcoil, to help identify the VLF signals correctly. VLF types have super-sensitivity and are not hindered by the presence of high mineralization. They will, however, produce a positive signal on some out-of-place magnetic ore stringers which contain different amounts of magnetite than that to which the instrument is "balanced" or adjusted. This kind of response generally happens only at very shallow depths and when the target is in close relation to the searchcoil. It is a result of the "near field effect" of the searchcoil. This slight disadvantage can be overcome readily by completely understanding VLF operation. The VLF types are the greatest boon to prospecting since the invention of explosives, but all manufacturers and detector dealers should explain the slight handicap thoroughly. Since the *DETECTOR OWNER'S FIELD MANUAL* covers the entire treasure hunting/prospecting field, it should be carried by anyone engaging in these activities.

tion a pencil at right angles on top of the ruler directly in the middle of figure six. You now have equal distance from each side of the pencil to each end of the ruler, approximately five-and-seven-eighths-inches. Consider the space occupied by the pencil as the null or silent area. (It represents also the dead center of tuning, with equal space allotted to each operating zone on your metal detector.) Consider the space between the pencil and the left end of the ruler as the "metal" zone; the distance between the pencil and the right end of the ruler, the mineral zone. (However, remember that on the detector the mineral zone may be silent, though for practical purposes the zone is still there, even if not responding audibly.) The distance from the pencil to either end of the ruler may be considered what is usually referred to in electronics as the "dynamic operating range." In other words, it represents the range or distance from where audible sound first starts (threshold) and increases to its fullest peak (saturation point).

With the ruler and pencil still in place, follow the manufacturer's instructions for tuning on any VLF type. You will be instructed to lower the searchcoil toward the ground. If the soil is negative (mineral) and the audible tone decreases in volume, the instructions will say to advance the terrain (ground or zero) control either clockwise or counterclockwise, dependent upon the way the control is wired. In making the adjustment for negative (mineral) you will ALWAYS advance the main (common) tuning control forward into the metal zone. Illustrate this procedure for yourself. Slide the pencil slightly to your left into the designated metal zone. Notice as you *continue* to adjust the tuning you change the dead center of tuning. The metal zone is becoming more narrow; the mineral zone (on your right) becomes longer or wider. This is always the procedure if the ground is negative (mineral). If the ground is HIGHLY mineralized the pencil (representing the null zone or dead center of tuning) may have to be moved to your left as far as numbers one or two on the ruler. Notice how this keeps decreasing the dynamic operating range of the "metal" zone and making the "mineral" side longer. This is necessary to zero or balance out the effects of the mineral. As the control (pencil — dead center of tuning) was advanced farther into the metal zone, the zone became more narrow and also more prone to overloading of the circuit.

Test your VLF type detector in this same manner. After you have advanced the ground adjustment and the tuning controls sufficiently to zero out the negative effect of highly mineralized ground (both controls have to be adjusted in a harmonious step-by-step procedure), test the detector adjustment with small coins or other metallic objects. Notice the objects respond correctly as metallic. Now obtain a sample of common rock with a *high* magnetic iron content. Test it in the same manner as a coin. If the ground control was forcefully advanced TOO far into the metal zone to eliminate the ground effects, THE COMMON ROCK WILL RESPOND FALSELY AS METAL. It did NOT overload the circuit as is so common on a standard TR type but truly responded positive (metallic). This "false" response was caused by the necessary shifting of the null zone (dead center of tuning) and the detector no longer has the ability to identify metal *vs.* mineral correctly.

The null zone remains stationary on both the BFO and standard TR types. If the standard TR searchcoil produced a uniform pattern it would also identify metal *vs.* mineral correctly, the same as does the BFO type. On the VLF types the null area (dead center of tuning) is shifted by the operator to allow the elimination of negative ground effects. This is what causes all VLF types to produce false "metallic" signals in certain areas and why VLF operation must be studied to identify metal and mineral correctly. These false signals are caused in certain areas that contain out-of-place iron mineral rocks. It is the peculiar iron content and composition of the iron ore that causes the false signals. Remember, however, those false signals are produced mainly on rocks buried at a relatively shallow depth. This peculiarity is discussed in more detail in other places in this book.

Let us discuss the use of the VLF type in a mineral salt area, ocean beaches, *etc.* (A place where the ground responds as positive or metallic on a BFO detector can safely be classified as positive or conductive.) Mineral salts when wet become conductive and respond as metal on all types of detectors except the PI and PRG. Simply reverse the tuning procedure for positive ground. Referring again to the pencil illustration, return the pencil to the center of the ruler. If the searchcoil responds to the ground as positive slide the pencil to your *right* into the mineral zone. The metal zone is enlarged as you continue to adjust the positive ground out. The so-called mineral zone becomes more narrow or shorter until the limit of adjustment is reached. On some ocean beaches the conductive salt content is so high that it is impossible for the VLF to tune out the positive ground completely. The situation will always depend, of course, on the conductive salt content on that particular beach and upon the operating range of the detector. Remember, when the VLF type was adjusted over highly mineralized ground that sometimes out-of-place rocks with a MUCH higher magnetic iron content than the surrounding area responded falsely as metal. Now that the detector is adjusted in the opposite zone (mineral side) to permit exclusion of the conductive salts, the detector reacts just the opposite, also. The slightest amount of magnetic iron now is violently negative. Caution must be exercised here to prevent the detector from drifting or slipping back into the null zone or you will miss some of the deeper targets. These actions of the VLF type detector may remind you of a standard TR type being used over highly mineralized ground in that it is difficult to keep the instrument tuned at a slight positive. This particular use of the VLF on wet mineral salts is satisfactory in most areas, but the VLF will not be nearly so efficient as when it is used on negative (mineralized) ground.

When you use the terrain (ground zero) control to adjust out either negative or positive background response, you also change the null zone (dead center of tuning). This is what causes the metal response from rocks of a much different magnetic iron composition than those on which the detector is adjusted to operate. Therefore, when you use the VLF type detectors in areas where digging may be difficult, such as inside mines, in rocky canyons or places where drilling and blasting might be necessary (places which probably contain mineralized hot spots of magnetic iron), you can carry a BFO type to identify possible false signals correctly. If the mineralized hot spot or magnetic vein is within the detection range of the BFO, the detector will identify it correctly as negative. If the BFO produces NO response, neither negative (mineral) nor positive (metallic), there is either a *false* signal (highly mineralized spot) or correct *metallic* response that is BEYOND THE DETECTION RANGE OF THE BFO.

This situation will hardly ever occur since almost all false responses on out-of-place hot spots are relatively close to the VLF searchcoil and within easy detection range of a BFO with large searchcoil. This is why I have attempted to explain in such detail the advantages and disadvantages of the new VLF type detectors. It is now easily possible for anyone to find conductive veins and high grade ore pockets in old abandoned mines (and working mines). The VLF type detector, properly understood and employed, will place electronic detection of precious metals forever on the map.

DETECTION OF GOLD IN MAGNETIC VEINS

Can it be done? Yes and no. Overly enthusiastic advertising and unqualified comments have created some confusion in this area in regard to the VLF detector. If you are searching in a mine with your VLF tuned (adjusted) to the matrix (the inside mine wall) and cross or pass over a magnetic vein that contains a deposit of conductive metal (gold, silver, *etc.*), you might detect the magnetic vein only as an out-of-place hot spot. You might even miss it entirely due to the fact that the content of one deposit electronically offsets or balances out the effect of the other. (It would depend on where the ground control is set in regard to the original null zone.) If you did locate the more highly mineralized area (magnetic vein) and adjusted the VLF to the magnetic content of that specific area, conducting your search following the magnetic vein, THEN you would detect the conductive deposit (gold, silver, *etc.*) WITH EASE. All these procedures are possible, and *highly productive* if you completely understand the VLF detector response in regard to these out-of-place hot spots that were not originally adjusted out by the terrain or zero control. Remember the terrain or zero (ground) control is continually shifting the center of tuning each time the detector is adjusted to a different mineral content. This fact must be taken into consideration when identifying a target. The very low frequency type TR will, however, some day *completely* replace the standard TR due to its greater depth of penetration and mineral-free operation. The BFO type detector can always however, be used to identify metal *vs.* mineral. As a team, the VLF and BFO detectors are unbeatable.

KINDS OF FERROUS IRON

This subject has always created confusion among even the most experienced detector operators. As far as the detector operator has been concerned, all types of ferrous iron are supposed to produce negative or mineral response. Positive responses from a target are supposed to indicate non-ferrous composition. Neither statement is true. To help eliminate confusion and prevent the opening of another barrel of snakes, I shall attempt to clarify the matter from my own experiences.

There is no such thing as ferrous *vs.* non-ferrous identification, except on sophisticated sensors that recognize all types of metals. ALL TYPES OF FERROUS IRON DO NOT PRODUCE MINERAL (negative) RESPONSE. Only one type does it predominantly — Fe_3O_4, magnetic iron oxides in the natural form. For example, hematite (Fe_2O_3) in its PURE FORM is actually conductive (the same as gold). It is non-magnetic and responds to a metal detector as metallic. Hematite crystals are actually used as diamond substitutes in jewelry-making. Hematite is rated over *six* in the hardness scale and is considered a gem stock by rockhounds. It is *very* difficult to locate in the United States in its pure form (before it became heated from volcanic action). There are great hills and deposits of hematite (in red volcanic form) in the U.S., but these deposits have been hot (received heat) and will respond as negative on your metal detector.

What does heat do to ferrous iron? The melting point of hematite, limonite and magnetite is 2750° F. When the hematite became hot, oxygen was introduced into its composition and it became magnetic. Practically all of the hematite masses in the U.S. have been hot and are magnetic to a certain degree. They also respond faintly as mineral (negative). The reason they do not respond violently as mineral (negative), as does magnetite, is that to some degree they still retain a conductive content which helps to cancel out part of the negative response. Look at a sample of hematite that has been only slightly hot. If it is of gemstone quality it will still be black and very hard. It will have an outer coating of iron that shows the bubbles or spots where the heat tried to penetrate. The outer layer is all that the detector sees and the instrument will show a slight negative reaction.

When you are gold panning or dredging, you will be attempting to separate the gold from the black sand (con-

The author is shown removing the concentrates from the sluice box into a pan for further removal of unwanted sands and gravel. After panning these concentrates, the fine gold will be transported home for additional refining or cleaning. Some may have to be reclaimed by the use of mercury. Mercury is NOT recommended for the concentrating of all gold recovered. Whether you remove the mercury by "burning" or retorting it coats the gold and considerably lowers its value for jewelry use. The small dredge being used is an Oregon Super Jet. It is quite portable and efficient but somewhat noisy due to its two-cycle operation. This type of small dredge is becoming popular for the weekend prospector due to its high maneuverability and suction power.

centrates). Actually the concentrates do not consist of PURE black sand. Part of the so-called black sand will be grey sand. This is hematite. You may use a magnet to pick up or sort out the magnetic black sand, but it will not get the non-magnetic grey sand. You could pour this grey sand on a piece of metal and heat it to 2750° F. after which the hematite would become magnetic and could be picked up easily with a magnet. After this sand becomes hot it becomes magnetic and becomes violently negative to your metal detector.

This is not to infer that the metal detector is a magnetometer; it is not. However, the ferrous iron takes on a different permeability effect in regard to its outer layer of composition which the detector sees as non-conductive. The detector then responds as to mineral. There are too many forms of ferrous iron to list here. Any good book on minerals will contain information on most of them and give their correct composition. On many others you will need to go to a library and dig into larger, more comprehensive volumes. Such information actually has NO bearing on operation of your detector other than that you will be helped to learn how to correctly identify metal *vs.* mineral with your detector.

Some manufacturers let their advertising on metal/mineral detectors get out of hand, resulting in confusion for both experienced and beginning prospectors as to what detectors will do and how they detect. The confusion is understandable as there were no reference books available in this field. Nowhere in all the available literature can you discover that NONE of the ferrous irons are magnetic, EXCEPT MAGNETITE, Fe_3O_4. Confusion has always reigned in regard to ferrous or non-ferrous, hard and soft metals. Many early day manufacturers and advertising writers continually made reference to "ferrous, non-ferrous," "hard and soft metal tuning," "responds to metal only, does not respond to mineral," "a metal detector, not a mineral detector." Regardless of all the suggestions from those of us with years of actual experience in the field, it was not until the 1960's that a knowledgeable electronic engineer began exposing and refuting these phrases in a series of articles for an Eastern publication related to treasure hunting. This man has great stature in the electronics field, and his articles and information finally succeeded in putting such misleading phrases to rest. ALL metal detectors are adversely affected by the mineral (negative) content of the ground, regardless of whether they are PI, PRG, BFO, TR or VLF. The VLF type is the least affected as you may adjust the effects out, but it is affected, nonetheless.

All that responds as metallic on a metal detector is not necessarily metal, just *conductive*. The mineral salts (ocean salts) are a good example of this. When they become wet, they become conductive. ALL types of detectors do NOT recognize them as conductive, but the average, everyday types that we use certainly do. This is just one example of something that responds as metal that is not metal, ONLY CONDUCTIVE. Another good example is pyrite . . . yes, iron PYRITE, "fool's gold." Obtain a good, high grade specimen of iron pyrite and test it on your BFO detector. If it is really high grade, it will probably even respond as metallic on your standard TR detector. If it is definitely positive (metallic), look it up in a good mineral identification book. You will find it described as follows: *NON-metallic;* brass-colored; metallic luster; hardness 6 up; specific gravity about 5; melting point about 1260° F.; chemical content FeS_2; composition iron sulphide; occurrence in veins; associate mineral (metals) gold and silver; remarks-source of sulphur. (Here is another example of a ferrous substance which is conductive, the same as the other non-magnetic ferrous irons.)

After you are convinced you have a positive specimen break off a small piece, place it on a metal object, and use a propane torch to heat the pyrite to its melting point. (Propane will do this on pyrite, but on other iron and gold you would need a higher heat.) After you have heated the pyrite specimen to the red-hot point, let it cool and then test it again on your detector. It will now respond as *negative* (mineral). Crush it up and place a magnet over the crushed particles. You will notice it has become magnetic. That is exactly what happens to other ferrous iron when you add oxygen to their chemical composition. None of this information is likely to make you any the richer, but perhaps it may help explain certain occurrences and what happens to rock in volcanic actions and heat zones.

GOLD, as well as many other metals, has been found

in volcanic rock. The metals simply became trapped there. They generally are encrusted with volcanic ash or rock on the outside and the only method of detection is by electronic means. Never pass a chance to test volcanic rocks, even if your partners insist it will do you no good. Many a tidy bundle has been made from this pursuit. With the advent of the new VLF type detectors that do not see the volcanic (iron) rock, I myself have great hope for productivity in this field. Many a nestegg has been hatched in old "worthless" rock piles.

These words about iron were not intended to bring out the geologist and mineralogist in you, only to explain my experiences and observations over the last thirty-some years with regard to different irons and detector responses. Perhaps this information will help provide some of you with many sunny days in pursuit of your hobby; perhaps it may explain away some untruths. Either way, I hope my experiences with minerals have helped you in some small way to understand your detector more fully as regards its responses to minerals.

NEGATIVE GROUND

This term would apply to any type of soil where the presence of iron oxide or magnetic iron in its natural state creates a negative mineralization. In other words, if you tune your detector in the metal mode, lower the searchcoil to the ground and notice a drop or slowing down of the signal, the ground contains negative mineralization. This effect is created by the presence of natural magnetic iron (Fe_3O_4, iron oxide) in its unprocessed state. If you were to tune your detector in the mineral mode of operation the signal would be positive, indicating mineral ground.

For all practical purposes in the operation of any type metal/mineral detector, the only MINERAL that will respond as mineral is non-conductive iron (black magnetic sand, magnetic iron in the natural state before smeltering). Gold, silver, copper and other natural non-ferrous metals are just that, natural metals. Remember, if the detector responds as MINERAL, all you have detected is the presence of natural magnetic iron. (A rusty tin can will cause the same response because the iron is returning to its natural state.) If your detector is tuned in the METAL mode of operation and the ground is highly mineralized or negative, you will find operation difficult. You will lose some depth because of the high mineralization. The black sand or magnetic iron does not stop the signal but, depending on the type detector, it does tend to absorb some of it.

The indication of magnetic iron detected in any mineral or rock sample does not mean that the sample does not contain gold or other precious metals, only that the sample contains more Fe_3O_4 than it does metal. Also, the indication of metal in any sample does not mean that it contains just metal, only that it contains more metal than it does mineral. This is, however, one of the very few tests on which you can positively rely. If a BFO detector is tuned in the proper METAL mode of operation and the indication is positive (or metallic) on any given sample, you can be assured the sample does contain some conductive substance in sufficient quantity to affect the electromagnetic field and produce a metallic signal. This is one of the great advantages in the use of a high quality BFO detector, especially for prospecting and mining. There is no reliable way to tell if you have found precious metal, but when a properly tuned BFO detector indicates *metal* you can be 100%-sure you have found some type of conductive substance.

The indication on an ore sample that gives no signal — neither metal nor mineral — is no indication that it contains neither metal nor mineral, only that the sample might be barren of both or that it may contain equal amounts of both as one could cancel the effect or presence of the other. This is highly unusual, but could happen. I realize some statements are repeated occasionally; however, I know some of you will skip around in this book and I want to be sure you do not miss these important facts.

The electromagnetic field created by the searchcoil will not respond to metals in many forms, some of which are precious metal in sulfides and soluble forms. It will respond only to metal in conductive form, so-called free milling ore, of sufficient conductivity to interrupt the electromagnetic field.

NEUTRAL GROUND

This applies to any type of soil or ground that has absolutely no effect on your metal detector, positive or negative. In other words, when you tune your detector in the normal mode of metal operation and place the searchcoil on the ground, it will neither speed up nor slow down. This is called neutral ground.

Sometimes the operator will decide the ground is relatively free of mineralized iron oxides, though this is not generally the case. The earth is like a huge ball, the center of which is thought to be molten lava. Entire continents contain some degree of magnetic iron in its natural state. This iron dust could have been placed either by natural fallout or from deterioration of the surrounding area.

Test this by scrubbing the soil with a small magnet. Inspect the magnet to see if it contains any small particles of black sand. If it does, the ground contains some degree of mineralization, but the negative effect of the mineralization has been canceled out by some type of metallic or conductive substance. This conductive substance is usually mineral SALT from the seas which has been spread over portions of the earth by tidal waves, winds and other acts of nature. You will notice most neutral ground occurs in coastal areas such as in the eastern United States. In some small areas of the western states you will find small, isolated areas of neutral ground. This has been caused by the dispersal of small particles of conductive ores or min-

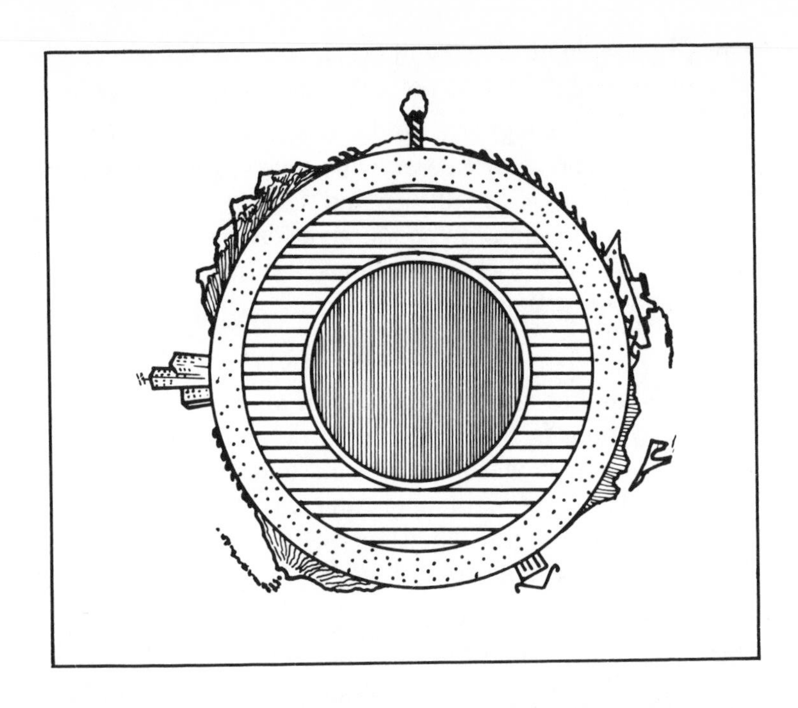

This is an illustration of a cross section of the earth. It shows how the entire surface (or somewhat below surface) is composed of iron. In most areas this iron is the negative mineral to which detectors respond.

eral salts in sufficient quantity to cancel or override the effects of the iron mineralization. This will also occur in some valleys below mining areas rich in metallic-type ores. In other areas the presence of clay that contains aluminum in some conductive degree tends to offset the effect of the natural iron in the soil. It is mostly unimportant to us why some ground is neutral. Just be thankful it is because it enables the operator and his instrument to operate successfully without bothersome negative soil.

POSITIVE GROUND

Positive ground will respond to the metallic mode of tuning on most detectors. That is, if you tune your unit in the normal metallic mode of operation and move the coil toward or contact the soil, you will receive a positive signal, similar to metallic responses. This does not mean the soil contains no mineral, only that the detector reads the soil as metallic. This happens because of many factors. Perhaps originally the soil had little erosion and it has received enough conductive metallic particles to override the magnetic iron present and create a positive reaction from your detector.

In the case of some seashores there may be large quantities of black sand or magnetic iron present, but, when you operate below the tide line while the sand is still wet from conductive sea water, you will probably get a positive or metallic reaction on your detector. You might move inland where the tide has not wet the soil and the ground there might respond as negative. This variance will occur on negative, neutral and positive beaches, depending on the amount of saturation received from conductive salts or the amount of mineralization present.

In prospecting you will discover a few mine tunnels that have enough naturally distributed low-grade metallic ore to cause the entire tunnel to read as metallic. When this occurs, the richer pockets will respond even more.

Below old mining locations in some highly mineralized areas, you will discover the entire ground reacts as metal. This is usually caused by the natural distribution of fine metallic particles from low-grade ore which are of sufficient quantity to cancel or override the effects of mineralization.

Positive ground provides problems to the metal detector operator, problems similar to operating around highly negative-type areas. You may still be successful . . . if you take your time.

TESTS will explain the terrain (be it negative, neutral or positive) and what it contains. You may use the hand magnet to scrub the soil to tell quickly if there is any black magnetic sand present or if the area is actually free from magnetic iron, an occurrence seldom found, however.

SECTION III

Prospecting and the Metal/Mineral Detector

While it is impossible to guarantee instant success in the prospecting field, if you follow three basic rules you can be almost certain to find *GOLD* and other precious metals in conductive forms. Non-conductive or highly mineralized pockets (predominantly Fe_3O_4—magnetic iron) containing rich but lesser amounts of precious metals may also be located provided they are out of place. Either way, it is almost impossible to fail if you persist in your search.

FIRST. Choose the correct TYPE of detector for prospecting. This does not necessarily mean some particular brand name or model.

SECOND. You must have patience. Learn to understand your detector fully and become proficient in its use.

THIRD. You must learn the correct place to start your search. No one finds gold or other precious metals where they don't exist. Stick to known, productive mining areas until you have become familiar with the telltale signs of mineral zones.

There are many things that can be done with a quality detector. It can open up completely new areas of action for the prospector. The thousands of modern-day prospectors who have been successful with the metal/mineral detector have employed the virtues of WISDOM and PATIENCE, plus a great deal of RESEARCH.

CHAPTER 15

Metal/Mineral Ore Sample Identification

METAL

Gold, silver, copper and other valued metals are natural non-ferrous metals which will respond to your detector as *metallic*, provided they are in a conductive form and in sufficient quantity to disturb the electromagnetic field of the searchcoil. Some high grade ores are in sulfides and other forms, and are NOT of sufficient conductivity to give a reading. Most free milling ores, however, that contain high grade metal in the conductive form will produce good response.

MINERAL

For all practical purposes, the only mineral that the metal detector recognizes as "mineral" is Fe_3O_4, magnetic iron oxides (in other words, magnetic iron ore or magnetic black sand). It is extremely simple to determine if the ore contains a predominance of either metal or mineral. If the specimen of ore contains *neither* metal nor mineral, your detector would produce no indication. There is, of course, a remote possibility the specimen may contain electrically *equal* and *exact* amounts of metal and mineral. In this case one would neutralize or balance out the effects of the other and NO indication would be received.

However, if the specimen reads as "mineral" this does not mean METAL is not present, only that there is a predominance of mineral. If the specimen reads as "metal" you can be certain that it contains metal in conductive form in such a quantity that you should thoroughly investigate that specimen. This factor makes the metal/mineral detector the most important tool of today's successful prospector and miner.

MODERN PROSPECTING. Notice two types of detectors are present, both the BFO and VLF types. These will generally take care of the requirements for all conditions and situations. The old burro is absent, but the four-wheeler is much more practical and comfortable!

BENCH TESTING ORE SAMPLES

Place your BFO detector prone on a table or bench, using one of the small searchcoils, preferably the 3½-inch or 5-inch. Tune the detector in the METAL mode of operation, adjusting the sound to a moderate beat to enable faint signals to be heard more clearly. (Be sure to remove all metallic objects from your hand; *i.e.*, rings, watches, *etc.*) Quickly move an ore sample toward and away from the center of the searchcoil. Test with the center of the coil as this area will give the best response.

If the sample contains NEITHER metal nor mineral, or has electrically equal amounts of both, you receive no response. If the sample has a predominance of metal in a detectable form, you will hear a slight beat and sound increase as the sample comes closer to the searchcoil. If you are using a detector model incorporating a sensitivity meter you will see a positive signal or pointer movement to the right, indicating the presence of metals. If the sound dies or the beat slows when the sample approaches the coil, the sample contains a predominance of mineral or natural magnetic iron (Fe_3O_4). This does not mean the sample contains no metal, only that the sample contains MORE mineral than it does metal.

One of the advantages of using the BFO all-purpose detector to test ore samples is that it will indicate the presence of METALS in any conductive form. That is, you can be absolutely certain whether the specimen contains METALLIC substance in some form or other.

Obtain samples of galena (lead), silver, gold ore and just plain rocks. By conducting your bench analysis you will become familiar with the type and amount of response to low grade and high grade ores. Also, many types of ore do NOT respond to metal detectors. Only those containing metal in conductive form in sufficient quantity to disturb the electromagnetic field will cause a detector to respond positively.

For example, a large garnet will respond as mineral (negative) on a sensitive BFO because the garnet contains enough magnetic iron to respond. When checking samples that have responded as metal, you will generally notice a metallic appearance on the inside. When samples that appear metallic but respond as mineral are sawed or slabbed, you will generally notice a streak of magnetic iron on the inside. Since there was sufficient iron to override the small amount of metal, the sample responded as mineral.

The newer VLF types have advanced to where they can be used to correctly identify metal *vs.* mineral, "out of place hot rocks," and perform all other prospecting operations. They are more difficult to learn to use, but the extra effort will pay off!

CHAPTER 16

Mines, Dumps and Veins

SEARCHING OLD MINE DUMPS

Working some of yesterday's forgotten ore dumps has become a very profitable national pastime. Some large mining companies were after only certain minerals or metals. The human eye could not see inside the ore, and many good pieces were discarded on the dump. The electronic metal detector can detect metals (of conductive form) inside almost any type of rock. Some dumps have been completely reworked for the minerals and metals that were left behind. Others are just awaiting the present-day prospector who is using a sensitive metal/mineral detector.

One of the greatest failings of the average treasure hunter who searches old mine dumps for high grade ore is to search the ore dump as one would in usual treasure hunting. The searcher simply expects the detector to respond to small (BUT RICH) amounts of metals among all the rubble of extreme MINERAL concentration. It is not reasonable to expect any detector — no matter how sensitive — to detect small specimens of metal in all the mass of rock that is generally very heavy with iron or that is sometimes a complete jumble of low grade metals which produces a background of metallic-type response.

When you encounter a dump that may contain high grade ore, lay your detector prone on the ground. Tune it in the metallic mode, using a small coil such as used in the bench analysis (3½-inch or 5-inch). Pick a few small samples of rock from the pile to test for metal content. If after a reasonable period of time you do not find any metallic indications, move to another area of the dump. During the working period of the mine perhaps there was only a certain portion of the dump that could have received the tailings from the vein. The rest may be only debris from

SEARCHING OLD MINE DUMPS. Many mine dumps, rich in metals and minerals, are awaiting the present-day prospector who uses a sensitive metal/mineral detector. The BFO is recommended because of its uniform searchcoil detection pattern and its continuous sound feature.

the mine's shafts and tunnels. Take rock or ore samples from many different locations, especially from the higher sections of the dump because this is where some of ALL the different pieces were dumped at one time or another. If after prolonged testing you do not recover at least a few metallic specimens, it may simply be a dump where the ore was of some type to which a metal/mineral detector could not respond. The ore may also have been of a very low grade composition of little value.

The best procedure is to do some research and pick areas that produced the free-milling type of HIGH GRADE ore which will respond positively to the detector. The pocket country (Washington State, Oregon . . . or almost every mining district, for example) is ideal for this type of searching. Most old timers merely wet the rock in order to see the gold and took only the high grade ore, commonly called jewelry ore. Jewelry ore is worth much more than the weight in gold for specimens. "High grade" is in great demand today, and I personally know some hobbyists who do quite well at this particular type of hunting. Someone else has already done the digging, and all the searcher has to do is to grade or analyze the discarded rock left on top. A good quality BFO metal/mineral detector will enable you to do this easily, provided the metallic content is of a conductive type and rich enough to respond.

Again I stress that millions of dollars have laid unnoticed on both small and large ore dumps, in plain sight of everyone. Absolutely anyone with reasonable ambition can use a good VLF or a BFO detector to excellent advantage on these discarded rock piles. You will be surprised at the valuable specimens you recover. Remember that a small rich specimen is worth many times its weight in gold or other precious metal. Obtain a metal/mineral detector with a small searchcoil and check a few small mine dumps or prospect holes. I will wager my entire life-span of fifty-odd years, plus a rep-

FIELD SEARCHING. There are many different facets of the prospecting hobby, such as looking for rich float, ore chutes or chimneys and deep ore veins.

utation for honesty and telling it like it is that you will recover many small, high grade specimens. Follow my instructions on ore sampling, and, if you are successful, pass the newly gained knowledge on to your friends. There are enough discarded ore dumps to furnish everyone with many days of pleasurable activity, not to mention the extra income that these dumps will provide.

SEARCHING OLD MINES AND MINE FLOORS

Old mines are no different from new mines except that the older tunnels and shafts were generally worked by less modern methods so that there is always the possibility of finding missed pockets or hidden veins. Great care should always be taken in the exploration of any abandoned shafts or mine tunnels. Many have become unsafe over the years and the danger factor is high. It is recommended that one always search such areas accompanied by a fellow hobbyist.

One of the most frequently overlooked yet most productive areas in any old mine is the tunnel floor. ALL the high grade ore had to come through the main tunnel before dumping or milling. It is almost impossible to move a large amount of rock, whether by ore cars or by hand, without dropping a few pieces. These small ore samples were generally covered by natural debris as the workmen drilled deeper into the earth. High grade ore has laid here unnoticed for many years, just waiting for the hobbyist who employs a quality detector.

Most inexperienced searchers merely enter the mine and quickly sweep or search the tunnel floor with the detector, much the same as they would conduct an outside search. This method is almost always unproductive because the fact has not been considered that the mine's mineralized background distorts any response. Use a quality detector to conduct a successful floor search. A VLF or a BFO is generally necessary because of the high mineralization and because you need not only wide tuning range but a correct metal/mineral response. Use a smaller-size coil, for example the 3½-inch to 8-inch. Place the detector in a prone position as in bench testing, and use your rock pick to dig under the top debris. Test small likely-looking samples. Move around to test many different spots, constantly keeping in mind that large high grade samples would have been seen by the original miner and ONLY smaller pieces would have been overlooked. However, due to the increased price of gold and the rarity of good specimens, this can be a very profitable venture. Old mine floors are very productive locations for small but valuable and maybe rare ore specimens.

If the metallic ore can be detected with a quality metal/mineral detector, you will probably receive a positive response, provided your detector is correctly tuned. However, you should save and later investigate any negative or high mineral indication samples. The ore may not respond in the metallic mode. If this happens the samples containing the pay streak will generally produce an unusual reading, perhaps more mineral content than the average rock. Any unusual piece of rock should be thoroughly investigated. The metal/mineral detector will not always be foolproof in this type of searching, but it will enable you to sort unusual and out-of-the-ordinary samples from the common rock.

SEARCHING FOR MISSED MINE POCKETS AND VEINS

We discussed pocket hunting in the search for surface ore pockets. However, pockets or small concentrations of ore occur at varying depths. Pockets or small ore bodies in some old mines have occasionally been missed only by inches. Vein locating also falls under this kind of searching as a vein is just a continuous streak of ore which sometimes pinches out or completely disappears. Careful searching with a metal/mineral detector does not guar-

SEARCHING OLD MINE FLOORS. One of the most productive but frequently overlooked areas in any old mine is the tunnel floor. Since all the high grade ore came out through the main tunnel, chances are good that valuable ore samples were dropped and covered with debris. These samples await the modern-day detector operator.

antee discovery of either a pocket or a vein. However, if you are experienced and the pocket or vein will respond to electronic detection, you have a good chance of rediscovery.

Searching old mines with a detector is probably the most underdeveloped and misunderstood hobby or profession in the world. This is because of the many failures encountered using early-day equipment and instruments that were not designed to operate inside mines or in highly mineralized areas. Underground searching differs slightly from searches conducted aboveground. Mine shafts and caves are often HIGHLY mineralized and require different methods and equipment. Your choice of detector types will include both the BFO and VLF. Other types can be considered ineffective, dependent on the mineralization present.

The VLF type is the best performer underground when high mineral content (magnetic iron) is present. Turn your VLF detector on; tune it to metal; hold it approximately four to twelve inches from the side wall (depending on the size of searchcoil used); and try to select a wide area where the response remains steady. According to the manufacturer's instructions, proceed to adjust the ground or terrain "zero" control to eliminate the background response. When this is accomplished, scan the walls and ceiling carefully, marking or taking note of any positive (metallic) response. Any negative response you hear can be disregarded. Ore containing a sufficient amount of conductivity will respond as positive (metal).

However, when the detector was adjusted to eliminate the effect of magnetic material, the center of tuning was also changed. In cases where the VLF type is adjusted to compensate for a high mineral background, the detector is caused to respond "positive" to isolated, out-of-place hot spots that contain MUCH MORE mineral (Fe_3O_4) content than surrounding material. This is one of the drawbacks of the VLF when tuning is adjusted to eliminate the effects of negative mineral. Any conductive (metallic) ore responds as metal, but it also responds as metal on any non-conductive hot spots that are isolated or out of place with respect to the area for which the tuning was originally set. At first this would seem to present an insurmountable problem. The VLF, however, can be adjusted to identify these hot spots correctly. If you wish you can use the standard BFO type detector, which has true metal *vs.* mineral identification, to determine whether the target is predominantly metal or mineral. The BFO will not penetrate the mineralized rock as deeply as the VLF, but, since most false indications of the VLF were from relatively shallow depths, it remains only for the BFO to identify conductive (metallic) ore correctly or to indicate just a spot containing a high content of magnetic iron. Though not foolproof, this method will eliminate most errors – AND, it eliminates much digging!

Rich ore will often be associated with magnetite. Sometimes it is possible to test a known piece of high grade ore from the mine and determine that it does contain valuable ore even though it also contains a large quantity of iron ore. If this is not possible and you are merely prospecting, take care to obtain different samples and determine their content by assay. After this procedure you will know what you are looking for and be able to recognize it based on the previous amount of detector response. No matter the content of the ore, any hobbyist who takes the time to experiment with his detector can spot unusual indications which warrant investigation.

The back-up detector for the VLF type can be a high-quality BFO which has shielded coils and Zero Drift features. Use a searchcoil of medium size, 12 inches or larger, as it will provide reasonable penetration and the ability to detect veins. Remember that the ore body is NOT solid metal so the response will always be rather faint unless the ore is of *extremely* high conductivity. To overcome the effects of the minerals present you will probably need to set the tuning slightly faster than normal. This presents no problem when using a good BFO with the continuous sound feature because you can hear all of the audio ups and downs. Move the searchcoil in a slow, continuous motion when sweeping mine walls. This method will tend to smooth out erratic behavior caused by the surrounding mineralization. Operate the searchcoil approximately four to twelve inches from the side wall or ceiling to obtain better penetration and identification under adverse mineralized conditions.

As stressed previously, ALWAYS investigate all large, unusual detector indications, whether metal or mineral. This cannot be overemphasized! Depending on the composition of the ore, the detector may respond either way. After you are familiar with the ore composition, you will be able to proceed with confidence.

Old mines have been woefully neglected over the years for several reasons. I predict this field of search activity alone will produce some of the most fantastic discoveries of the century when modern, quality instruments are employed. The searching of old mines can be rather frustrating because of the erratic behavior of many detectors when used inside mine tunnels or caves. Patience and experience with VLF and BFO instruments will, however, produce phenomenal results. Just think of the HIGH GRADE ore pockets and VEINS that have been missed by ONLY INCHES!

DEEP VEINS

A myriad of deep, well-hidden veins crisscross the earth at various intervals. Many are rich in metal and mineral content; others completely lack any value whatever. This perhaps is one of the toughest and most unproductive phases of the prospecting hobby. The vein is sometimes TOO deep for detection. It may respond either as metallic or mineral, and one has no way of knowing until he has samples to test. However, the rewards of persistence may be very handsome indeed. These aboveground searches may be conducted with either the VLF, two-box RF or BFO type.

My good friend, Frank Mellish of London, England, scans this seam for traces of "missed" crystalline wire gold. This location in the Cascade range of Central Washington is the only known area in the world producing true crystalline wire gold. You can see the seam where gold has already been taken out. You need a good BFO or, better yet, one of the newer VLF's to find this gold and then the mass of the pocket must be of considerable size. When selecting one of the VLF's, be sure to select one that has been proved in this kind of work.

The vein may be metallic and respond as metal, or the vein may have a predominance of iron oxide (Fe_3O_4) and respond as mineral. Regardless of the detector response you receive, evaluate it by considering its magnitude, in which direction and how far it runs. Within reason, you may judge the depth by the amount of response. This will not always provide an accurate gauge, but drilling or digging may be warranted if the width and length of the response area are unusual.

Set the tuning at a medium beat or sound and make wide sweeps or swings with your searchcoil. The response area may be quite wide, and if you fail to cover enough area in your sweep you may not be able to determine the edge (or start and finish) of the signal. Deep veins are usually a composition of several metals and minerals; therefore, extreme caution must be observed when listening for signals as most will be very faint.

One of the best locations for *easy* vein searching is around the construction of new roads. The logging roads of the Northwest have exposed many veins and small stringers. Watch the deep pass cuts when traveling highways. Perhaps the workmen did not recognize the ore or mineralization for what it really was. A quality BFO detector will quickly test these locations. It remains only to gather and test samples. Over the years many samples of semi-precious stones have been gathered by rockhounds from these very locations. It stands to reason that if the workmen missed these they may have missed some of the metallic and mineral samples that are MUCH harder to recognize.

Researching most mining districts will reveal that veins generally run in the same direction. This enables you to conduct your search in a definite pattern so you have the best chance to bisect or cut the vein. Veins have a habit of disappearing and then reappearing in the strangest places at the most inopportune times. One of the saddest and oft-told stories of the West is ". . . the vein pinched out . . . we lost the vein . . . it just disappeared." When you recount all the lost veins that contained fabulous riches, you realize the value of the metal/mineral detector.

The VLF type used with the largest searchcoils available will be one of the surest producers in the future relocation of lost veins. The ability to penetrate mineralized background will permit explorations in areas *never before possible.* The modern prospector who employs this new detector type will certainly have more than an even break. Past failures in searching for lost veins have made many professional miners wary of the electronic detector, but there is NO question that the VLF type with *mineral-free operation* will give renewed hope to the faithful and help to produce wealth for many. Did you ever wish you could see into the mine wall for just one more foot? Think of the hard work you could have saved! Consider the shafts and tunnels that honeycomb areas that were rich producers in the old days. How many times the shaft or crosscut missed the vein or pocket by only inches! Any prospector who fails to recognize the possibilities is passing up the easiest strike in his life.

SEARCHING FOR MAGNETIC OR NON-MAGNETIC VEINS. One of the best locations for *easy* vein searching is around the construction of new roads and highways. The author is attempting to determine the magnetic content (non-conductive) or metallic content (conductive) of a small vein that was exposed by road construction in the high mountain country. Numerous readings were received in the area when the deep-seeking VLF detectors were employed.

CHAPTER 17

Field Searching for Precious Metals

The phrase "field search" includes many meanings and covers many different facets of the prospecting hobby, such as looking for rich float, ore chutes or chimneys, and deep veins.

For example, take the search for high grade float . . . small pieces of ore that have broken off from the main ore body and been carried by water across the surrounding terrain. There are many ways in which pieces of ore can be displaced by either nature or man. A miner might carry a piece of high grade ore for years and finally misplace or lose it. Sometimes the finding of such a piece of ore will result in the search for ore bodies not even in the vicinity of the find. Float usually moves with gravity downhill or downstream, depending on nature's quirks. When a piece of high grade float is found, it is best to try to determine from which watershed it came. Your first concern should be to search for nearby additional pieces to ascertain in which direction to start.

Conduct your search in gullies, creek bottoms and other terrain where you believe the heavy float had to stop. Logical reasoning will dictate where to search. Remember that the small pieces of rock are heavy and gravity eventually places them in the lowest place. Search uphill from your first find. While this isn't always correct, it is a good bet. If you're fortunate enough to find another piece you know you are on the right track to the mother lode.

Your best choice of instrument, because of the continuous sound factor, is a high quality BFO detector. Of course, success depends on whether the ore sample contains a metal of a conductive nature to which the detector will respond. As you sweep the searchcoil in a regular search pattern, you will notice that almost all of the rocks are mineralized. When your detector is tuned in the proper metallic mode of operation, the rocks normally found in most areas will respond as mineral. If the sound or beat frequency INCREASES as you pass your searchcoil over a rock, you have found a specimen that contains metal which is generally non-ferrous, meaning precious metals or other substance of conductive composition.

The best coil sizes for searching for small mineral or metal float range from the 3½-inch to 5-inch to no larger than the 6-inch. These sizes will give better-than-average ground coverage and are still small (HOT) enough to respond to metallic float. Pieces of float generally are not pure or solid metal and will not give a loud signal, as would a tin can.

Tune the detector at a speed determined by the mineralized background. If it is highly mineralized, set the tuning at a fast beating sound and try to cover as much ground as possible without being careless. This method usually produces the best results, and you will be able to cover low spots and dry gullies quickly.

Definitely save all rocks that respond as positive or metallic. Also, if you notice one that shows TOO much mineral content, bring it in for later analyzing or assaying. ALWAYS INVESTIGATE THE UNUSUAL. With luck and patience your search may be rewarded with riches or with, at least, an interesting discovery.

ORE CHIMNEYS AND POCKETS

An ore chimney is a spot where volcanic pressure has pushed material containing metals and minerals upward through a fissure or crack in the bed rock. The material cooled as it reached the surface and formed a small, isolated pocket of ore. Searching for this type of formation is called POCKET HUNTING. The pocket usually has been covered by gradual erosion and fallout, but never very deeply. Over many thousands of years the ore has gradually decomposed, and gravity has carried the decomposed ore downhill. Sometimes this ore will be small pieces of rich float or placer gold. If the gold is concentrated enough and relatively free of iron oxides, it has a good chance of responding to a sensitive metal/mineral detector.

POCKET HUNTING

One of the most famous states for pocket hunting is Oregon. The pockets generally occur on hillsides where gradual erosion has caused the placer gold to be carried downhill. To locate the pocket old prospectors followed these traces uphill by patient panning. Digging down a few feet, they removed and hand-sorted the rich ore, taking only the rock with *exposed gold.* (Remember this when you search old mine dumps.) The holes left are relatively shallow and, although there is not too much rock left by which to locate the dump, there are plenty of these small mines worth highgrading.

The location of rich pockets is accomplished by using a VLF or a BFO detector which employs a medium-to-large-sized coil. You need the depth of the larger coils, and they will also speed your search by covering more ground. Tune the detector in the metallic mode of operation and sweep the searchcoil approximately four to six inches above the ground. Listen closely for the rather faint indication you will receive from a pocket. The signal is faint because the pocket is covered by a few feet of other material and the gold ore is NOT pure metal. Nevertheless, as mentioned earlier, if the pocket is relatively free from iron oxides you will probably get a positive, or metallic, response. Pockets, as a general rule, are not very large. The faint audio response tends to remind one of the detector signals produced on small concentrations of birdshot, such as found at a shooting range. (The metal detector is used to locate HOT or concentrated spots where shot has landed, and reclaiming devices are then used to harvest the used pellets.)

POCKET HUNTING. Rich ore pockets occur in almost every mining district. For fast ground coverage and depth penetration use a BFO which employs a medium to large-sized coil.

Some of Oregon's pocket gold occurs in porphyry. This makes digging easy as most of the pockets are relatively shallow. The gold is easy to remove by crushing the porphyry and panning the concentrates. Actually, gold occurs in no set pattern. It is best to remember the old saying, "Gold is where you find it."

Rich pockets occur in almost every mining district. Washington State has probably the rarest types of gold to occur in pockets. One type is the famed crystalline wire gold that has been found around the town of Liberty, Washington, located in the Cascade range of mountains that extends from Canada to the Oregon coast. The crystalline gold nuggets are beautiful beyond expression and are found only in this mountain range. This is probably the most valuable gold ever found. Should you ever want to purchase one of these specimens, start with the story where the man once said, if you have to ask "how much" then forget it; it is too expensive for you! I can state without reservation that the person who locates one of these pockets will probably be able to goldplate his faithful detector and retire for life. With a good quality BFO metal/mineral detector, pocket hunting can be a rewarding experience.

CHAPTER 18

Nugget Hunting

NUGGET SEARCHING IN WATER

The ability of metal/mineral detectors to locate small gold nuggets in mineralized stream beds has been over-advertised and oversimplified to the point where people think all they have to do is purchase a detector and head for the hills. Presto! Instant riches! Nothing could be further from the truth. Nevertheless, by following proved guidelines and paying close attention to the type of detector you choose, it is definitely possible to find the *larger* gold nuggets where they do exist. Actually, many detector operators do better at this activity than dredging operators and with far less work and expense.

Search methods in water and in highly mineralized mountain streams differ greatly from the methods used on dry land. Most important, you need a detector that will operate adequately among mineralized rocks and over uneven terrain. It is absolutely ESSENTIAL that the searchcoil be FULLY ONE-HUNDRED-PERCENT FARADAY-SHIELDED. Otherwise, the motion of the running water as it flows over the searchcoil will generate static and false signals, making the detection of small nuggets practically impossible. Next, the detector must have HIGH sensitivity to small metallic objects such as small gold nuggets. A searchcoil ranging in size from three inches to no larger than eight inches is preferred. Your choice of detector types should be made between the BFO or VLF. The standard TR type is not recommended because of its quick-response operating characteristic. The VLF type must have total ground cancellation features to be effective.

If your choice is the BFO detector, it should be able to accommodate small nugget coils or probes. It should be extremely stable and employ the Zero Drift feature. The

The author is trying to locate gold nuggets in New Mexico. This type of searching generally results in finding many small, worthless metallic objects that have washed downstream over the years. However, IF there are any gold nuggets of detectable size and IF the operator has patience and perseverance, his VLF type detector will certainly locate them. One of the tremendous advantages of this type detector is that it is perfectly suited to working in highly mineralized areas where super-sensitive response is necessary. As this new type of detector becomes more widely known, nugget hunting will become extremely popular and you will hear many tales of fantastic success.

surface rocks and deeper bedrock are generally highly mineralized in nugget country, and the constant sound feature of the BFO detector is extremely helpful. The signals will be driven "up and down" by the mineralization, and the constant sound will enable you to distinguish the sharp "ZIP" or increase in sound when you detect a nugget.

You should choose slightly smaller coils than those used on dry land because of the maneuverability required among the rocks and boulders. The coils must be small enough to have high sensitivity on extremely small metallic objects, yet still have size enough to gain the necessary depth. These requirements narrow the choice down to the 3-inch nugget probes in most cases, especially the probes manufactured for BFO detectors. Because of their construction they will operate more reliably in high mineralization areas than will larger coils. They are wound in configurations such that they will read primarily on the bottom of the coil. One-inch or ¾-inch probes would not cover enough area, and the extra loss in depth caused by use of smaller probes would not be acceptable.

Tuning speed should be governed by the amount of mineralization present. If there is HIGH mineralization, simply increase the setting (tune the detector at a FASTER speed). If there is normal or low mineral content, tune the detector at a more moderate speed to achieve better performance.

You will experience considerable difficulty when using the BFO in certain rocky stream beds that contain a high amount of mineralization (magnetic iron). The uneven surface and extreme negative effect of black magnetic sand and rocks make the recovery of small nuggets sometimes impossible.

Nugget searching in this type of terrain is best accomplished by the use of detectors featuring complete mineral-free operation. The VLF types designed and engineered in the EXTREME low frequency range (one kilocycle-plus) will best accomplish this. These detectors will penetrate magnetic black sand perfectly. They operate on the magnetic phase principle which completely eliminates the negative nature of the mineralization. Depending on the operator's skill, large gold nuggets can now be found under mineralized rocks and in black sand quite successfully. The specially designed VLF types are advanced design, super-sensitive instruments and should NOT be confused with the popular standard coin hunting TR (IB) detectors.

To eliminate the negative effect of the highly mineralized stream bed, adjust the VLF according to the manufacturer's instructions. Operate the searchcoil about

UNDERWATER NUGGET HUNTING. By following proved guidelines and with careful searching where gold nuggets do occur, it is possible to find and recover rare gold specimens. A BFO detector with minimum drift, hot sensitivity and wide tuning range is successful in most areas. The VLF type is required in more highly mineralized streams.

one to four inches aboveground, moving it slowly over the search area. Many manufacturers provide smaller searchcoils for nugget hunting, but careful field testing will prove the standard 8-inch size to be the most practical for the VLF type, *even on the smaller nuggets.* Earphones may be necessary if background noise (such as running water) distorts the audible response.

Carry a plastic "Gravity Trap" gold pan, plus a small garden trowel or shovel. When you think a metallic target has been indicated, slip the shovel carefully under the spot and lift the contents into the plastic gold pan. Test the pan with your detector to see if the detected object is in it. (A plastic pan is necessary because a steel pan would interfere with the detector's response.) If the metallic target is in the pan sort the gravel carefully and try visual recovery. If this is impossible try panning the contents in water. Perhaps the target is only a small piece of ferrous trash or it could be a hot rock with more mineral (iron) content than that for which the detector was adjusted. Either way, this procedure is a quick, easy way to find the larger gold nuggets.

The same method of recovery may be used on old dredge tailings. Many dredges used a small trommel screen (revolving perforated drum) to separate the smaller rocks and nuggets. Sometimes larger nuggets grizzly off (are discarded) with the rocks. Since this was a frequent occurrence, many dredge tailings may be profitably worked. Picnickers in Washington State near the old town of Liberty have found many such large nuggets among the discarded tailings. You can see some of these fabulous specimens displayed by many banks in that immediate area. Any large nugget from one ounce up is easily located. Even the small five-pennyweight size presents no problem to the experienced operator who is using one of the new VLF types. (Of course, small nuggets cannot be detected at any great depth.)

Practice by placing small metallic objects or small

pieces of large copper wire among the rocks and natural gravity traps in the stream. Sometimes a hunter's spent bullet will set your heart to pounding. But, again, it can be the real thing . . . gold!

An unbelievable amount of small metallic junk abounds in isolated mountain streams, as you will discover. It seems almost impossible that man could have been in these places in such numbers and lost so many different things. As with gold, small metallic objects are heavy and eventually wind up in a streambed, carried there either by rains or gravity. Most, however, have been thrown into the water. The knowledge of natural processes can be in your favor as you realize that small, heavy objects eventually wind up in a streambed. A body of water that has passed through mineral areas has the best possibility of containing some of the riches.

Many different type detectors are advertised for "nugget shooting," and this sometimes confuses the buyer. Only two types currently available to the public perform this function efficiently: the BFO and the Very Low Frequency (VLF). The BFO responds well to nuggets in most cases and also locates black sand pockets with ease. However, the Very Low Frequency (VLF) type penetrates black sand better than a BFO, and when properly tuned it is not adversely affected by mineral background. If your local dealer does not have the proper type detector I suggest you contact a major manufacturer who produces both BFO and VLF types and take his suggestions. However, as with the purchase of any specialized equipment, I recommend testing before buying, if possible. Sometimes a short trip or telephone call to someone who is experienced with different types of detectors can save many hours of unproductive labor.

The higher cost of specialized detectors for nugget searching will probably be forgotten in the elation of success. Many years of frustration and failure have haunted the nugget hunter. Nevertheless, with the advent of the specially constructed low frequency VLF's complete with *total* elimination of negative ground effects and the very recent improvements in BFO instrumentation that have considerably lessened the effects of negative or mineralized ground, you can rest assured many tales of success will be forthcoming. In the years to come nugget shooting is certain to become one of the most productive forms of prospecting. The cost of specialized equipment becomes unimportant when you consider that two small nuggets will pay for the instrument. Just remember, you will not get apples unless you get into an apple orchard! The same is true of gold. You must get out and look for it where it has been or might be found and use the right *type* equipment.

NUGGET HUNTING IN DRY PLACER DIGGIN'S AND DRY WASHES

Hunting for nuggets in old placer diggings and in the bottom of dry washes is probably the most productive and rewarding phase of prospecting. In remote desert areas where water has never been available and the only method of recovery was "dry panning" or "dry washing," there are untold millions-of-dollars-worth of small nuggets just lying on the surface. They rarely are detectable by eyesight, but they lie often at very shallow depths, almost in plain view. Investigation of low-lying areas with a metal/mineral detector can be very rewarding. There are knowledgeable nugget hunters who have done quite well by working dry or desert areas in highly mineralized locations. Stories have been published over the years of authenticated finds of small, wheat-grain-sized to exceptionally-large specimens, and the use of the metal/mineral detector is practically the only method of locating such deposits. It is certainly one of the fastest.

For nugget searching in dry areas your choice of detectors should include two general types: the BFO or the VLF type that is built especially for penetration of black sand (mineralized soil). The BFO will perform quite well in dry areas and has the advantage of being able to locate and correctly identify black sand pockets while nugget hunting. The mineral-free operation VLF type detector is not a metal/mineral detector and *does not* locate black sand pockets, but it will produce phenomenal results on nuggets large enough to respond. The smaller searchcoils are best suited for location of small metallic objects. You will be able to check the bottoms of dry washes, arroyos, ancient streambeds quickly and even huge flat areas with relative ease and speed. Do not expect great depth on these small nuggets as they generally lie among mineralized rocks, a situation which presents a problem for even the best quality detector. Of course, there is always the possibility of discovering a LARGE unusual nugget; many such have been found and are in museums or banks.

With each search sweep you are covering much more ground than a man could shovel, even if you obtain only surface readings or perhaps a depth of only one inch. All things considered, you will be surprised at the actual amount of ground you can pan or work. This is the tremendous advantage of the use of the metal/mineral detector for prospecting. The use of the plastic "Gravity Trap" gold pan is almost mandatory here also. You will need a plastic gold pan capable of panning dry materials, and the "Gravity Trap" pan uses the same riffles as the standard dry washer. Follow the same recovery instructions as for nugget hunting in streams. Follow the instructions included with the pan for successful dry panning.

If your choice of detector is the BFO and you wish to search for nuggets, tune your detector in the metallic mode of operation. This makes it extremely simple for the BFO to distinguish small concentrations of black magnetic sand. When the detector is tuned in the proper metal mode, a small nugget will produce a "zip" or slight increase in the beats. When you pass over a small pocket of

NUGGET HUNTING IN DRY PLACER DIGGIN'S AND DRY WASHES. This form of nugget hunting is probably the most productive and rewarding phase of prospecting. Searching old dredge tailings can also be very profitable. Advanced BFO and VLF detector instrumentation offers unlimited opportunities in the nugget hunting field.

magnetic black sand, the sound will almost completely decrease or at least appreciably slow down. All BFO detectors have the constant, continuous beating sound. This operating characteristic will enable you to check *all* indications without being forced to tune in the mineral mode just to find the mineralized pockets. Black sand pockets will sometimes contain a concentration of placer gold. Your tuning speed should always be governed by the amount of mineralization present. In highly negative or mineralized ground, simply increase the speed. For neutral ground or ground relatively free of mineralization set the tuning at a more moderate beat to obtain the best results.

Operate the searchcoil as closely to the ground as you can to receive both the best indication on these small nuggets and a slight advantage in depth. Govern your swing or sweep according to the situation. Remember to move slowly if you have the tuning set at a slow motorboating sound. If you have it set rather fast, then you can sweep faster without fear of missing the audio signal. Practice will determine the best tuning speed for each situation.

If your detector choice is one of the many new VLF types you may simply adjust the ground or terrain control to the area of search and operate the coil as closely to the ground as possible. This will gain all depth possible on the smaller nuggets because of no interference from the mineralized soil. It is not practical to locate black sand pockets simultaneously, as with the BFO, but if the area is heavily mineralized the VLF types will produce the best results. The specially designed VLF's will penetrate black sand with ease, and fantastic results have been reported in many areas. Search loops should be in the medium 8-inch size to facilitate ease of movement among the boulders or rocks. Only metallic objects of sufficient conductivity and a few out-of-place mineralized rocks will read on this type detector, and black sand presents absolutely no loss in depth penetration. This instrument will completely revolutionize the mining industry. The search for high grade conductive ore (gold, silver, copper, *etc.*) is now possible with the advent of these very low frequency (VLF) types that see through a jumble of mineralization (iron, Fe_3O_4). This will certainly make feasible the reopening or re-exploring of many abandoned mines and other areas.

I cannot overstress the following point: all detectors are not suitable for prospecting. The poor results received from some detectors have simply caused many an individual to quit this interesting and profitable hobby. Careful attention to selection of a detector will enable you to recover many items of interest from the desert or dry areas. Research will prove quite helpful here, and careful decisions as to search area can be very rewarding. Bonanza or mother lode areas that have produced unusually large nuggets or high grade ore (free milling type) are almost certain areas of success. Regardless of where you live — north, south, east or west — gold has been and continues to be found practically everywhere. Again, gold is where you find it. At today's inflationary prices it takes only a small amount to add up to a nice nest egg. The system of electronic detection is bound to produce great riches for many individuals over the next few years. Dry placer areas were almost completely untouched by the old miners, and the cost of a good detector is nothing in comparison to a few small gold nuggets.

CHAPTER 19

Black Sand Pocket Locating/ Underwater Dredging

Concentrations of black sand do not necessarily contain gold, but the magnetic sand and the gold are both heavy and they tend to stop in nature's natural traps. When you are operating your sluice box, notice that the black sand also traps behind the riffles the same as in nature's sluice boxes, the rocks and crevices in running streams. Locating black sand pockets with a metal/mineral detector is sometimes possible both under water and on dry land.

Your choice of detector should be the all-purpose BFO. An absolutely *drift-free* detector is a "must" in cold mountain streams. The temperature factor was always one of the treasure hunter's major stumbling blocks before the invention of Zero Drift circuits and searchcoils.

Your choice of size in searchcoils will differ slightly, depending on whether you would like to detect metallic gold nuggets, as well as black sand pockets that might contain gold. It is possible to compromise and do both with reasonable efficiency. In underwater searches for black sand pockets the coil size choice should be governed by the nature of the bottom of the stream or river. If the bottom is rocky and you have to search among large boulders, the smaller size (3 to 6-inch) will be more practical. If the bottom area is smooth and sandy and you wish to get as much depth as possible, choose the 12-inch or larger searchcoils for maximum depth.

When using the BFO detector to seek black sand pockets, set the tuning at a moderate beat. This slightly slower beat will enable you to distinguish the response more clearly. Smaller searchcoils operate best tuned in the metal mode. The beat will slow down as you pass over concentrations of black magnetic sand. However, if you

Tommy T. Long uses a "Gravity Trap" gold pan to clean concentrates recovered from the small gold dredge. Allan Cannon tries his luck at locating black sand concentrations for later application of the suction type dredge. Allan uses a BFO type detector, the only type that can identify metal *vs.* mineral correctly. Locating the black sand may be hampered by a high mineralization content in the area and may even be virtually impossible. It is still, however, worth the small effort which must be expended and is highly successful in many dredging areas. Notice the bedrock that is visible. Such an area is generally the most profitable and the easiest from which to obtain gold with minimum effort. A suction dredge, BFO type metal detector and gold pan present the perfect combination for success.

pass over a large metallic GOLD nugget, the beat will speed up. This characteristic provides a double capability so that with practice you will be able to distinguish a black sand pocket while having the opportunity to locate an *unusually large* gold nugget or metal object.

If you use the larger searchcoils (12-inch or larger), it is advisable to tune in the *mineral* mode of operation. (These coil sizes are such that the detection of small metallic objects like gold nuggets would be almost impossible.) Tuned in the mineral mode, the beat would increase as you pass the mineralized pocket of black sand. Using the

There it is . . . perhaps a pocket of black sand that might contain a large amount of gold! Fred Heine, co-owner and designer of the lightweight, 4-inch Super Jet gold dredge, is seen in background. Notice the new discharge design and the twin 15-inch sluice boxes. That is a lot of water for a 4-inch to handle! BFO detectors featuring Zero Drift are used by many dredgers to assist in locating mineralized pockets. You cannot always find the pockets and they may contain no gold — but it's far better than guessing. For purchase or dealership information on this fantastic new "gravel hog" gold dredge write or telephone OREGON GOLD DREDGE, LTD., 120 Monroe Street, Eugene, Oregon 97402, 503-343-6741.

mineral mode with the larger coils also allows faster ground coverage and more depth.

This same method may be used in dry land searches. Keep in mind that the coil size will govern whether you discover an unusually large nugget while conducting the black sand search. Also, remember that the BFO detector will give *true* readings on either black sand pockets or conductive metallic objects. Testing of various types of detectors will quickly confirm the BFO to be the most practical choice for this type of prospecting.

Transmitter-receiver detectors (including even the

especially-built, very low frequency, VLF; the radio-frequency RF two-box; the induction balance, IB; pulse induction; discriminator; *etc.*) have many uses, but correct identification of metal *vs.* mineral is not one of them.

When two or more operators are using a metal/mineral detector in conjunction with an underwater suction dredge, *sometimes* it is possible to work the more heavily concentrated areas. One person can operate the detector and the other, the dredge. This is no sure-fire method as conditions will vary, but it is always worth the

The largest inland suction type dredge being operated in the continental United States. Owned and operated by John Mock, Lucile, Idaho, it is an 8-inch suction dredge operating at approximately 25-35 feet in depth on the famous Salmon River ("River of No Return") at Lucile. Approximately 4-6 experienced divers are employed. John is one of the few successful gold divers who has truly made it pay. Large nuggets are quite common here. John and this large dredging operation have been pictured in many magazines over the years. BFO type detectors were experimented with to locate black sand pockets, but because of the extreme depth of the overburden the operation was not successful. BFO detectors will aid most operations in magnetic pockets in either wet or dry locations, but depth of detection and an irregular river bottom will influence the outcome considerably. Thank you, John, for this picture of your fantastic dredging operation.

effort because it is better than a shot in the dark. Success will depend on the condition of the stream bed and whether it is possible to operate among all the mineralized boulders. Small creeks in certain areas are profitable producers if this system of detection is used. Many rivers have bottoms that permit detectors to be employed with some success. Here again, this is no certain method of location, but many small dredge operators have found it successful in a few areas. I feel it is worth the small investment of a good BFO detector, but success is not so certain as in other areas of prospecting.

CHAPTER 20

Rocks, Gems, Minerals and the Metal/Mineral Detector

Without question, the most important and useful tool of the rockhound (besides his faithful rock hammer and patience) is the metal/mineral detector. Properly used it can be very rewarding, *but* it should not be used as the ultimate answer to the positive identification of all minerals and gems. Nothing will ever replace knowledge gained from experience in the identification of semi-precious stones and gems. The metal/mineral detector should be used as an added accessory to the rockhound's field equipment. It will aid in the location of many conductive metallic specimens the human eye cannot distinguish or identify. Many high grade specimens of different ores can be overlooked on any given field trip. While the human eye cannot see inside an ore specimen, a good quality detector can.

"Metal" is defined as any metallic substance of a conductive nature in sufficient quantity to disturb the electromagnetic field of the searchcoil. Gold, silver, copper and all the non-ferrous metals are just that — metals. "Minerals" refers only to minerals which respond to a metal/mineral detector. Detectable minerals are primarily magnetic iron and iron oxide. The proper chemical content is Fe_3O_4. (Refer to the preceding information on metal and mineral identification.) What it all boils down to is: if the BFO detector responds as "metallic," bring the target in for it contains conductive metal in some form. If the target responds as "mineral," this indicates only that the specimen contains *more* mineral than it does metal in any detectable form that would react to the detector. This means that for a few minutes work you just might wind up with a high grade metallic sample that has been passed over for years by your fellow rockhounds.

The author holds a high grade piece of pyrite (FeS_2) removed from a small deposit located while nugget hunting with the BFO type detector. The pyrite is NOT classed as a metal, but it is conductive and responds the same as other natural non-ferrous metals. This point has created confusion in the mind of prospectors and is best explained as follows. ALL metallics in a predominant (basically, more "positive" metal than "negative" non-conductive natural iron) and conductive form respond as metal. ALL magnetic minerals respond as negative (non-conductive) but some minerals (salts) when wet will then respond as conductive. Also, any type mineral that will conduct electricity will respond as "metallic." However, as stated throughout this book, these are rare occurrences and need not concern the average, everyday operator. Stick to the relevant facts and most common occurrences. If a target responds as metallic, bring it in for later identification. It is usually valuable.

FIELD SEARCHING FOR GEMS. Pay close attention to old mine tailings when searching for gems. Thousands of rockhounds search dredge tailings and ore dumps with beat frequency oscillator detectors. Many small but valuable discarded gem specimens now may be found with electronic instruments. Rich ore that is conductive is also easily identified with these instruments.

Conduct bench tests to familiarize yourself with the responses produced by both metal and mineral. Use specimens with which you are already familiar for this will aid you greatly in your future testing of samples. Refer to the bench tests conducted for the identification of ore samples. When conducting your field search, use the detector as an *aid*, not as a complete searching tool. In other words, test any likely-appearing rocks. This kind of testing and investigation will vastly increase your knowledge and may produce for you that valuable specimen desired by all.

In the identification of all mineral *vs.* metallic ores, you will find that certain VLF and BFO detectors will produce the best results. This is in no way meant to downgrade TR detectors. The BFO and certain VLF's will *not* produce different responses with reference to the searchcoil area that comes into contact with the sample, as will the TR. Many rockhounds have turned from the TR metal/mineral detector, primarily because of the unreliable identification of known ore specimens. This operating characteristic is not a failing of the detector. The TR is merely the wrong detector choice.

The TR detector tuned in the metal mode will give different signals on the same specimen of metallic ore, depending on which part of the searchcoil the specimen touches. A large gold nugget placed on the receiver portion of the coil will produce a metallic or positive response. Place the nugget on the portion of the coil containing the transmitting coil and it will respond as "mineral" or negative. Place a small pebble of extremely high grade magnetic iron ore on the receiver coil and it will respond as "mineral." Place it on the transmitting portion of the coil, and it will respond as "metallic" or positive.

Always choose quality detectors when testing ore samples because the entire surface of the coil will respond the same, either metallic or mineral. Proper attention to the type of metal/mineral detector used as an aid in identification of mineral or metallic ores will result in the full enjoyment of your hobby and lessen the chances of false identification.

Pay close attention to old mine tailings when searching for gems. There could be a high grade specimen of metallic ore around such a location and the metal/mineral detector used as your extra eye will possibly identify it quickly. Certain gems such as the "thunder egg," have a coating of magnetic iron. Some forms of jade and even garnet respond to the mineral side of a quality, sensitive detector. Experiment! A whole new world will open up when you become familiar with your detector. You will be able to spot check any promising-looking rocks, and perhaps you may find that "worked out area" isn't so barren after all.

Your detector dealer may have a wide selection of known, easily recognized gem or ore samples. Ask him to demonstrate various marginal samples on different types of detectors. The results of such testing will enable you to enjoy your hobby more profitably. If the dealer does not understand identification procedures with a metal/mineral detector I suggest you recommend he read a copy of this book. The possibilities of success with a good metal/mineral detector should not be brushed aside. There are countless thousands who employ detectors in their search for missed or unrecognizable gem stones. Many of these contain precious metallics and are worth small fortunes at today's prices.

Recently I attended one of the larger gem shows. A lady rockhound from Colorado was displaying and selling rock specimens from a well-known gold mine. She had picked the specimens from the discarded ore dump, broken them into baseball size, and priced them at $3. A few felt rather heavy, and I became curious as to gold content. I requested permission to test them with my BFO metal detector. Three of the specimens responded like almost solid metal. I explained to her that I thought they had a large gold content. She replied they were still for sale and she did not believe in detectors. I promptly bought them and had a rockhound friend saw them into slabs. Two of them were heavily loaded with gold, and some of the slabs later brought $125 each for jewelry making. I personally have high graded many "for sale" silver specimens and chosen the loaded ones. Perhaps this will give you some ideas for future use of your metal detector. Even a child can accomplish this testing with a BFO type detector. Many people are doing it. Think on it.

CHAPTER 21

Gold Panning and the Metal/ Mineral Detector

The metal/mineral detector is an aid to the gold pan? Yes, the metal detector can definitely be of assistance in locating pockets of black sand and may also be used to locate large nuggets of pure gold which is of a conductive nature. You will almost always have to make use of your gold pan to pan or sort through the rubble of rock and sand to locate the small metallic object that responded to the metal detector. The target may be only a spent bullet or

After you have concentrated the heavier material, whether by wet or dry panning, it is best to bring the concentrates home for slower, more careful panning procedures. Pan slowly into another pan so as not to lose any of the color into the larger container. It would take considerably longer to repan the larger container than it would the second pan. The author is shown here carefully separating concentrates recovered from the Salmon River in Idaho. Heavy small broken particles of garnet make this repanning necessarily slow. Gold is often difficult to separate from heavy concentrates of magnetite, hematite, garnets, *etc.* The "Gravity Trap" riffles are indeed an asset.

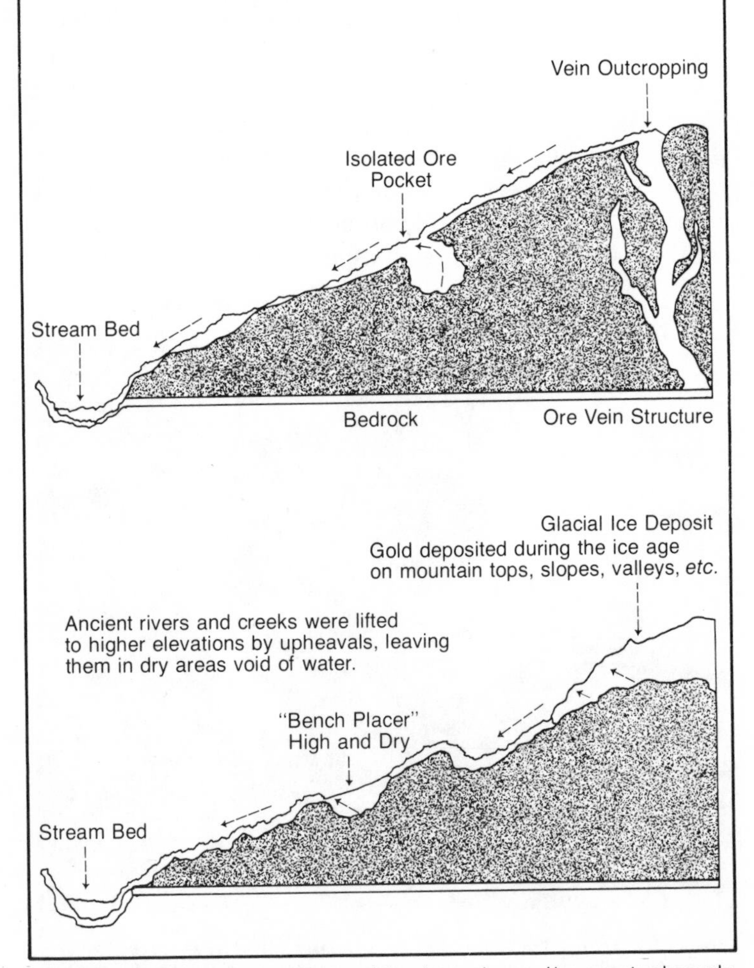

Unless the surface structure of the earth has been changed by recent upheavals, mountain slides, torrential downpours and so on, it is possible to follow the faint indications of gold to the source. Generally, this task is accomplished by patience, panning in grid patterns (made difficult by the absence of water), or visual searching. More recently accomplishment of the procedure has been made possible by newer types of electronic detection equipment. These dry areas hold great promise for the metal/mineral detector user.

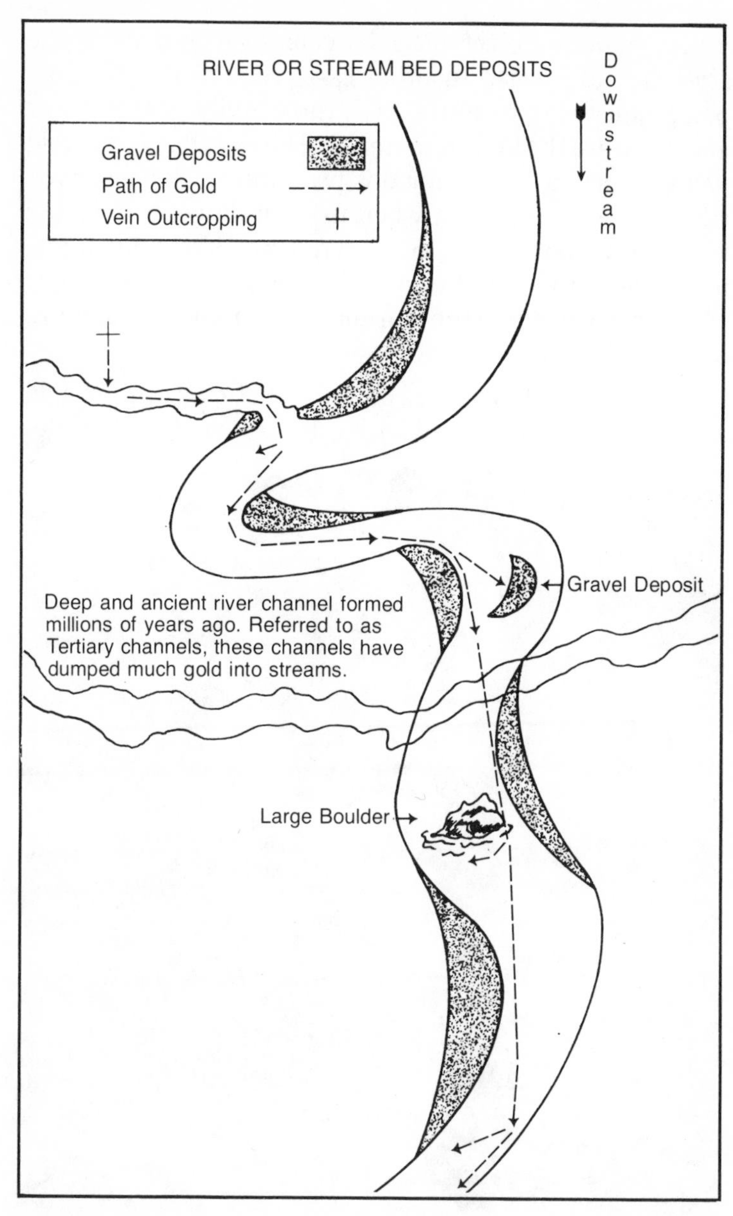

Regardless of the appearance of the present river channel and velocity of water flow, many changes may have taken place over the millions of years. River bends, current changes, amount of water flow and turbulence may have moved or displaced gold into unsuspected areas. Note the dotted lines that MAY indicate what *could* have happened. *It pays to look!*

other metallic object placed by nature or man in the stream or dry wash. In either case, you will find the gold pan a quick way to settle ALL the heavier concentrates and separate the lighter material. The use of a plastic gold pan is almost a "must" for this procedure. Obviously you cannot use a metal pan when using a detector over the pan. With a plastic pan you can quickly determine if you have placed the target into the pan on your first attempt. If not, you can dump the pan and dig more deeply or more carefully on your next attempt.

Location of black sand deposits that may contain gold can best be accomplished with a BFO type detector. Be-

The plastic "Gravity Trap" gold pan and the BFO metal/mineral detector go hand-in-hand. The "Gravity Trap" gold pan is recommended because it has sharp 90-degree riffles. Both wet and dry panning methods are possible. Follow the manufacturer's panning instructions for best results.

The author is shown dry panning on a high mountain glacier deposit where water is not available. Numerous responses produced by a VLF type detector resulted only in the recovery of two bullets (probably from a deer hunter's gun) and three extremely "hot" pebbles that were foreign to the area. The "Gravity Trap" pan is quite helpful. It may be used in dry areas to reduce the gravel and sand down to heavy concentrates quickly. Visual detection of the target, either large nugget or metallic trash, is then greatly facilitated. The plastic construction also makes the use of detectors possible while the target is in the pan.

cause the continuous sound factor and the availability of entire mineral or metal tuning, the BFO is quite well-suited for combination searching for BOTH nuggets and black sand deposits in a single operation. Due to the TR's quick response and non-uniform searchcoil configuration, the TR detector is almost totally impossible to use in heavily mineralized areas. However, if you are searching for large nuggets ONLY and the location is in a highly mineralized creek bed that abounds with rocks of high mineral content, the specially constructed VLF type (actually a TR) will definitely be the best instrument to use. You will lose the ability to detect mineral pockets but gain the advantage of super-sensitivity on small metallic objects (nuggets).

Regardless of the type detector you use, tune it according to the manufacturer's recommendations. Select the stream bed, wet or dry, and proceed either up or down stream, paying close attention to faint signals as the nuggets are generally rather small or may be deep. When you receive a target response, attempt to slip a shovel or like tool under the spot where the signal occurred. Be extremely careful as the small but heavier metallic objects always tend to sink into the rocks and will quickly become lost. Place the small amount of gravel or sand you scoop up into the plastic gold pan and check the material with your detector. If the target is in the pan the detector will respond. If there is no response, dump the pan and try to get under the object on the next attempt. After a small

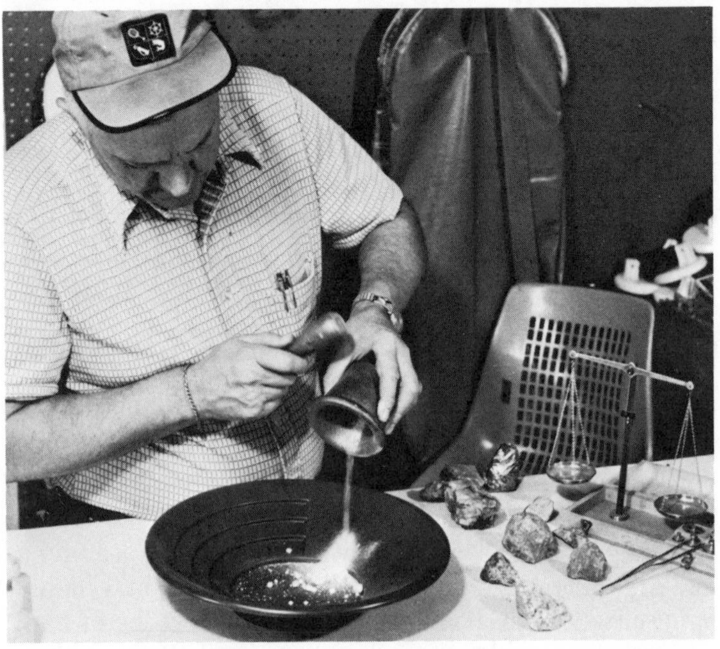

The author has mortared down a metallic specimen to determine its content. All metallic type samples do not contain values, but the odds are definitely in your favor that you will find value in the majority of the "positive" reading samples that your BFO tells you to take home. The fine dust from the specimen will be carefully panned, weighed and then catalogued as to weight and content. This method will provide only an approximation as to value but it is the quickest and easiest without a confirmed assay report.

amount of practice you will become quite proficient in this procedure and learn to save considerable time by inventing special tools for the job.

If you are searching dry wash or old placer diggings the procedures will be the same. The only difference is that the object will be easier to locate.

When searching old dredge tailings, you will lose many metallic targets on your first attempt at digging them out. They easily fall or work themselves down through the loose material. Once lost, they are very difficult to recover because they drop further on down through the dredge tailings, becoming almost impossible to relocate. In searching dredge tailings the specially constructed VLF mineral-free operation type detector produces the most satisfactory results.

In the future this type of prospecting and recovery will be one of the most lucrative and productive activities in the metal detecting field. There are many recent improvements in metal detectors, and when this fact becomes well-known to the public there will be another gold rush that could equal that of the '49'ers. The advantage of the modern searcher is that someone else "done the digging" . . . but missed some of the goodies!

You can add to your skills by attending a prospecting school. One such school is Lucky Leo's Prospecting School, 709 West Columbia, Pasco, Washington 99301. A. J. "Bedrock" Haley, owner of the school, teaches all facets of prospecting, including electronic prospecting. Write to "Bedrock" for full details.

Pictured below are contestants in one of the numerous state and national contests sponsored by GOLD PROSPECTORS ASSOCIATION OF AMERICA. GPAA conducts continuous instruction seminars throughout the U.S., teaching modern-day panning methods and demonstrating the latest equipment for the recreational miner and treasure hunter. Reprinted courtesy GPAA.

RULES OF THE CONTEST

PANS: Each contestant will start with a Garrett Positive "Gravity Trap" Gold Pan provided by the contest committee.

GOLD NUGGETS: Eight pea-sized gold nuggets, provided by the contest committee, will be placed on the sand, then the judge will push them into the sand to his first knuckle.

PROCEDURE: Beginning with pan of nuggets and sand, each contestant must remove ALL material (sand) except nuggets by panning.

TIMING: Timing begins when the panner touches the sides of the pan. Timing ends when the panner yells "GOLD." There will be 3 timers.

JUDGING: After the contestant indicates he or she has finished, a judge will count the number of nuggets and insure that there is no sand remaining in the pan. A ten (10) second penalty will be assessed for each lost nugget. There is a limit of 2 minutes.

TOP TEN PRELIMINARY WINNERS will pan in the Championship pan-offs.

CASH PRIZES will be computed from the entry fees to this contest and put into a pool.

1st place — Trophy & 50% of cash pool;
2nd place — Trophy & 20% of cash pool;
3rd place — Trophy & 10% of cash pool;
4th to 10th place will split remainder of pool.

THERE WILL BE A PRACTICE AREA SET UP FOR CONTESTANTS

The winner of GPAA's Oregon State Championship will automatically be entered in The Oregon State Open Championship in Baker this fall.

Columbia Basin Chapter #4 members, Natalie and Frank Meckle, were lucky enough to attend The Oregon State Planning Championship at Baker last fall. Participants were fourteen finalists from all over the state and included one woman and one teenaged young man. The winner was Jack Roberts, who also won the 1975 National Championship.

First Prize was a $189 Metal Detector, and all contestants were given the pan that they used.

For more information on membership and a FREE copy of their prospecting and treasure hunting publication, write to . . . GOLD PROSPECTORS ASSOCIATION OF AMERICA, P. O. Box 507, Bonsall, California 92003.

SECTION IV

Conclusion

I believe it is my duty to offer my apologies to any manufacturer or metal detector dealer that I may have unintentionally slighted in discussions of detector types. I assure you there was no such intention as it is my hope that the *DETECTOR OWNER'S FIELD MANUAL* will help the responsible manufacturer and dedicated detector dealer.

I feel, however, that my first responsibility lies with the detector operator who is performing the actual task of treasure hunting and also with the prospective buyer who has to thread his way, like I did many years ago, through the many myths and untruths that abound in some detector advertising. *I have paid my dues*, and now consider it my duty to help others in a manner consistent with good business practices. I know many professionals and technicians will gain some knowledge from this *MANUAL*, as was intended. This *MANUAL* will also help the beginner to protect his time and strengthen his efforts if he will follow the guidelines set forth.

You should practice with various detector types. You will learn that each has its good points. With your own detector, follow the instructions contained in this field manual and practice until you have mastered your detector. Strive to understand its good and bad points and to become a successful treasure hunter. Remember, that regardless of whether you ever become money wealthy, wealth can also be knowledge gained.

I have not made millions in this business, but for sure my experience lends 100% truth to the statement that I am regarded as one of the foremost "in the field" authorities on metal detectors in the U.S. I have cut money with many pro's over the years and all consider my word better than money. In my thirty-odd years of treasure hunting I have put most makes, models and types of detector instrumentation through every situation imaginable. I hope I have been able to impart some knowledge and trustworthiness to the business where it will do the most good.

Thank you.

Roy Lagal

INDEX

Metal Detector Dealers

The following is a list of reputable metal detector dealers throughout the United States. This is, by no means, a complete list nor does the omission of a dealer's name indicate that he is not reliable.

ALABAMA

HINKLES
4340 Pinson Parkway B-25
BIRMINGHAM 35215
(205-854-9621)

JOHN G. LINK
P.O. Box 682
310 Colonial
FLORENCE 35630
(205-766-0087)

ALABAMA TREASURE HUNTER
909 Chatterson Road
HUNTSVILLE 35802
(205-881-7772)

CONFEDERATE ORDNANCE
P.O. Box 66075
2202 Government Blvd.
MOBILE 36606
(205-473-3731)

BROOK'S CUSTOM DETECTORS
1026 Biscayne Drive
MONTGOMERY 36116
(205-281-1806)

ALASKA

NELSON ENTERPRISES
P.O. Box 814
KODIAK 99615
(907-486-3672)

ARIZONA

SIERRA TREASURES
2118 N. 4th Street
FLAGSTAFF 86001
(602-774-6512)

LUCKY TREASURE WORLD
6005-D West Thomas
PHOENIX 85033
(602-247-4506)

THE NATIONAL TREASURE HUNTERS LEAGUE
1309 West 21st Street
TEMPE 85282
(602-968-9295)

THE TREASURE SHACK
2190 E. Apache
TEMPE 85281
(602-968-0783)

MOREY DETECTOR SALES
3825 E. Hardy Drive
TUCSON 85716
(602-793-0071)

ARKANSAS

W. W. MOSLEY
P.O. Box 7
CAMDEN 71701
(501-836-5314)

OZARK TREASURE HUNTER LEAGUE
Industrial Park Rd.
HARRISON 72601

BILL'S DETECTORS
5623 R Street
LITTLE ROCK 72207
(501-666-6355)

TRAMMELL'S
619 Baker Street
MOUNTAIN HOME 72653
(501-425-3615)

L. L. LINCOLN
Route 1, 158 Pyramid Drive
ROGERS 72756
(501-636-6867)

ORCHARD'S METAL DETECTORS
Fishermens Choice
6 Mi. From Hot Springs on
Hwy 270 West
ROYAL 71968

CALIFORNIA

COE PROSPECTOR SUPPLY
9264 Katella
ANAHEIM 92804
(714-995-1703)

LO SIERRA MINING EQUIPMENT
534 Grass Valley Hwy.
AUBURN 95603
(916-823-1880)

C & J DETECTOR SALES
3104 Pepper Tree Lane
BAKERSFIELD 93309
(805-832-6619)

PROSPECTOR SUPPLIES
868 Ironwood Avenue
BLOOMINGTON 92316
(714-823-6165)

BREA BICYCLE & SPORTING GOODS
141 S. Brea Blvd.
BREA 92621
(714-529-3353)

AURORA PROSPECTING SUPPLY
6286 Beach Blvd.
BUENA PARK 90621
(714-521-6321)

RENCHER WELDING & MACHINE WORKS
560 S. Third Street
CHOWCHILLA 93610
(209-665-4219)

FRIEND'S TREASURE OUTPOST
Jct. Valiera & Fairfax Avenue
Route 2, Box 79-B
DOS PALOS 93620

THOMAS MURRY
P.O. Box 406
6001 Pleasant Valley Road
EL DORADO 95623
(916-622-5245)

ROY GENE ROLLS
Hwy. 32 at Sugar Pine
FOREST RANCH 95942
(916-342-4829)

FRESNO HOBBY CRAFTS
3026 N. Cedar
FRESNO 93703
(209-226-4880)

SOUTH BAY COINS
639 Ninth Street
IMPERIAL BEACH 92032
(714-423-2551)

FUMBLE FINGERS
1027 Brown Avenue
LAFAYETTE 94549
(415-284-7406)

ANTELOPE ACRES MARKET
Ron Farrell
48011 90th Street West
LANCASTER 93534
(805-948-4190, 942-7165)

GOLD NUGGET MINER'S SUPPLY
1302-9th Street
MODESTO 95354
(209-524-5822)

THE COIN SHOP
1516 Third Street
NAPA 94558
(707-255-8166)

KEENE ENGINEERING, INC.
9330 Corbin
NORTHRIDGE 91324
(213-993-0411)

PIONEER RECOVERIES
3510 Audubon Pl.
RIVERSIDE 92501
(714-682-4302)

BILL & MELBA DIBBLE
8851 E. Lansford Street
ROSEMEAD 91770
(213-287-7996)

B. C. DOUGLASS
1537 Placer Way
SALINAS 93906
(408-449-1131)

DENNIS E. WITKOWSKY
Coins and Supplies
P.O. Box 772
SAN BRUNO 94066
(415-589-8179)

GEM & TREASURE HUNTING ASSOCIATION
2493 San Diego Avenue
SAN DIEGO 92110
(714-297-2672)
(Closed Monday & Tuesday)

ARTS & HOBBIES
12323 Forest Trail
SAN FERNANDO (Lakeview Terrace) 91342
(213-899-1997)

MINING & LAPIDARY
131 10th Street
SAN FRANCISCO 94103
(415-626-6016)

JOHNNY'S METAL DETECTORS
209 N. Broadway
SANTA MARIA 93454
(805-922-8703)

PRICE'S TREASURES
P.O. Box 201
SHANDON 93461
(805-238-6487)

HIDDEN ROD SHOP
2623 Gardenia Avenue
SIGNAL HILL 90806
(213-427-8060)

GEMSTONE EQUIPMENT
480 E. Easy Street
SIMI VALLEY 93065
(213-348-6807)

COLORADO

THE PROSPECTORS CACHE
59 W. Girard
ENGLEWOOD 80110
(303-781-8787)

REG'S ELECTRONICS
4 Bonita
PUEBLO 81005
(303-561-3036)

C & D DETECTION ENTERPRISES
6195 W. 38th Avenue
WHEATRIDGE 80033
(303-424-7780)

CONNECTICUT

EDWARD PERCHALUK
304 Circle Drive
STRATFORD 06497
(203-378-1660)

J & E ENTERPRISES
1242 South Street — Route 75
SUFFIELD 06078
(203-668-0029)

FLORIDA

LAWSON STUDIO
1503 East Las Olas Blvd.
FT. LAUDERDALE 33301
(305-463-5311)

JAMES R. FORD
TREASURE CHEST
528 N. Eglin Pky.
FT. WALTON BEACH 32548
(904-863-1595)

AMERICAN INTERNATIONAL
P.O. Box 2186
HIALEAH 33012
(305-821-1500)

HARRY'S PAWN SHOP
519 Main Street
JACKSONVILLE 32202
(904-353-6971, 353-0320)

ROBERT TWAIT
OLDS KINGS ROAD TREASURE INN
6946 Old Kings Road S.
JACKSONVILLE 32217
(904-733-1928)

PLAZA CARDS & GIFTS
713 N. 14th
LEESBURG 32748
(904-787-4661)

KELLYCO
215 E. Horatio Avenue
MAITLAND 32751
(305-645-1332)

ZEPHYR TREASURES
2898 Zephyr Lane
MELBOURNE 32935
(305-254-2796)

MAIL ORDER ELECTRONICS
200 Mustang Way 13-B
P.O. Box 1133
MERRITT ISLAND 32952
(305-452-8236)

SEATECH METAL LOCATORS
985 N.W. 95th Street
Miami, FL 33150
(305-693-1431)

JOSH WILSON'S DETECTOR SALES
4704 N.E. 17th Avenue
OAKLAND PARK 33334
(305-776-1076)

TWELFTH AVENUE DRUGS
2435 N. 12th Avenue
PENSACOLA 32503
(904-433-6563)

CARL ANDERSON
Box 13441
TAMPA 33611

TREASURE SHACK
3934 Britton Plaza
TAMPA 33611
(813-833-9841)

GEORGIA

SOUTHEASTERN TREASURE HUNTERS
985 Woodland Avenue S.E.
ATLANTA 30316
(404-627-6019)

RICHARD K. JONES
FINDERS COMPANY
225 Upland Road
DECATUR 30030
(404-377-0974)
(Call Evenings)

ERNEST M. ANDREWS
Atlanta Tri-City Area
2755 Sylvan Road
EAST POINT 30344
(404-766-8141)

HOBBY SHACK
Zayre Plaza
834 N. Houston Road
WARNER ROBINS 31093
(912-923-6159, 922-2630)

J. C. BALLENTINE
P.O. Box 761
Hatcher Point Mall
WAYCROSS 31501
(912-285-3250)

IDAHO

OUTDOOR HOBBY SUPPLY
TOMMIE T. LONG
311 North 23rd Street
BOISE 83706
(208-336-8583)

Q'S TROPHY CABIN
3940 Overland
BOISE 83705
(208-342-5566)

ROY LAGAL
OUTDOOR HOBBY SUPPLY
2416½ E. Main
LEWISTON 83501
(208-743-1768)

POWERS CANDY CO.
POWERS HOME GAMES & HOBBIES
602 S. 1st Avenue
POCATELLO 83201
(208-232-1693)

ILLINOIS

RENE'S TREASURE TROVE
214 East Front Street
BLOOMINGTON 61701
(309-829-4538, 829-4058)

JERRY'S TREASURE
HUNTER'S SUPPLY
RR #1, Meents Lane
CHEBANSE 60922
(815-939-3815)

HARRY'S TREASURE SHACK
322 W. State Street
CHERRY VALLEY 61016
(815-332-5157)

GEORGE BEARDSLEE
1719 N. Finney Street
CHILLICOTHE 61523
(309-274-2649)

DETECTORS UNLIMITED
1671 Summit Street
GALESBURG 61401
(309-342-4032)

ELECTRONIC EXPLORATION
575 W. Harrison Rd.
LOMBARD 60148
(312-620-0618)

HIDDEN TREASURE
REV. JOHN J. COSTAS
3116 11th Avenue "A"
MOLINE 61265
(309-797-3098)

DEE'S BEAUTY SHOP
206 Reservoir Road
PEKIN 61554
(309-346-4377)

MID-WEST TREASURE DETECTORS
507 So. 8th Street
QUINCY 62301
(217-223-4757)

GARY & KAREN BENNETT
Indian Trail Road
VERONA 60479
(815-942-5290)

TOM'S POOL CENTER, INC.
801 North Green Bay Road
WAUKEGAN 60085
(312-244-4505)

MEMORY HOUSE
1 N. Chestnut Street
WEDRON 60557
(815-434-3568)

INDIANA

PAT'S METAL DETECTORS
R.R. #7, Box 145
ANDERSON 46011
(317-378-0475)

O-D WESTERN STORE
ROBERT A. EVERETT
RR #5
DECATUR 46733
(219-724-2097)

A-Z COINS & STAMPS
Glenbrook Center
4201 Coldwater Road
FORT WAYNE 46805
(219-483-3743)

J & J COINS
7019 Calumet Avenue
HAMMOND 46324
(219-932-5818)

L & M SALES
7310 Hazelwood Avenue
INDIANAPOLIS 46260
(317-255-4236)

PIONEER METAL DETECTOR SALES
10338 Pendleton Pike
OAKLANDON 46236
(317-823-4202, 898-4510)

WRAY'S TREASURE SHOP
RR #5
SEYMOUR 47274
(812-497-2537)

ALBERTSON'S SPORT SHOP
U.S. 30 East
WARSAW 46580
(219-267-3891)

IOWA

RICHARD CROSS
314 S. Main
BAXTER 50028
(515-227-3391)

CEDAR RAPIDS LOCK & KEY SERVICE
3217 First Avenue S.E.
CEDAR RAPIDS 52402
(319-365-5162)

NORMAN TRESLAN CONSTRUCTION
4 N. 16th St.
CLEARLAKE 50428
(515-357-2255)

HERB DUNN, JR.
METAL DETECTOR SALES
Route 4
INDIANOLA 50125
(515-961-4341)

McGREW OIL CO.
120 W. 4th Street
TAMA 52339
(515-484-2946, 489-2396)

DEAN BOYD
1047 Evergreen
WATERLOO 50701
(319-232-9484)

KANSAS

CARL CLARE
911 3rd Avenue
DODGE CITY 67801
(316-225-4701, 225-5005)

RADIO SHACK ASSOCIATE STORE
2609 Anderson Avenue
MANHATTAN 66502
(913-539-6151)

EPP'S COIN SHOP
112 S. Main Street
PRATT 67124
(316-672-6181, 6277)

SWAIM ELECTRONICS
1430 E. Douglas
WICHITA 67214
(316-262-0077)

KENTUCKY

GAMBILL LOCKSMITHING
1004 Comanche Ct.
ASHLAND 41101
(606-325-7931)

A. F. WALLER
P. O. BOX 72083
LOUISVILLE 40272
(502-937-8008)

PAUL PHILLIPS
109 Lake Street
NICHOLASVILLE 40356
(606-885-3648)

LOUISIANA

J & F ENTERPRISES
12211 Greenwell
Springs Road
BATON ROUGE 70814
(504-272-8500)

A-ABLE TREASURE ELECTRONICS
Route 1, Box 56 MN
BENTON 71006
(318-965-0277)

HAMMETT & SON ENTERPRISES
Route 1, Box 90
DELHI 71232
(318-878-2105)

HENRY L. MONTEGUT
437 Aurora Avenue
METAIRIE 70005
(504-834-2378)

MARYLAND

FRANK'S DETECTORS OF
GLENBURNIE
408 Arbor Drive
GLENBURNIE 21061
(301-768-3157)

SOMCO MACHINE CO.
Route 1, Box 272
WESTOVER 21871
(301-651-1516, 651-3964)

MASSACHUSETTS

FOUND ENTERPRISES
133 Prospect Street
AUBURN 01501
(617-832-3721)

LARRY VIOLETTE
Box 74
REHOBOTH 02769
(617-252-4497)

A. J. DUMAIS
DUMAIS ELECTRONICS CORPORATION
37 Spring Street
W. SPRINGFIELD 01089
(413-733-9548)

MICHIGAN

FAMILY FUN DETECTORS
9991 Harbor Beach Ct.
ANCHORVILLE 48004
(313-725-4682)

LLOYD R. BUZZARD
1724 E. Salzburg Road
BAY CITY 48706
(517-684-4765)

ROYSTON'S ROCK SHOP
7027 W. Vienna Road
CLIO 48420
(313-639-7070)

HUFFMASTER ELECTRONICS
1537 Monroe
DEARBORN 48124
(313-278-7922, 278-1940)

GRANT'S BOOK STORE
601 Bridge Street NW
GRAND RAPIDS 49504
(616-458-6580)

FINDERS KEEPERS
METAL DETECTORS
2112 Cumberland Road
LANSING 48906
(517-321-6594, 323-4250)

OLD PROSPECTOR'S SHACK
7007 Cooley Lake Road
UNION LAKE 48085
(313-363-7328)

TREASURE HUNTER'S SUPPLY
3930 Burlingame SW
WYOMING 49509
(616-538-1957)

MINNESOTA

MID-WEST METAL DETECTORS
8338 Pillsbury Avenue So.
BLOOMINGTON 55420
(612-881-5254)

G & P SPORTS SHOP
301 Elmwood Road
HOYT LAKES 55750
(218-225-2296)

MINNESOTA PROSPECTORS SUPPLY
Formerly of Red Wing, MN
902 Goodrich
ST. PAUL 55105
(612-226-5118)

MISSISSIPPI

K & Y KRAFT COMPANY
903 West 7th Street
CORINTH 38834
(601-286-8096)

EAGLE ARMS CO.
3115 Terry Road
JACKSON 39212
(601-373-4557)

HOBBIES UNLIMITED
P. O. Box 1161
1219 Nelle Street
TUPELO 38801
(601-842-6031)

MISSOURI

BLUE SPRINGS DETECTOR SALES
1012 Woodlynne Drive
BLUE SPRINGS 64015
(816-229-1559)

TWIN CITY GUN & CB
500 Cooper Street
CALIFORNIA 65018
(314-796-2166)

THE PROSPECTOR'S SHACK
975 Grenoble Lane
FLORISSANT 63033
(314-837-4703)

E & R DETECTOR SALES
P. O. Box 213
HILLSBORO 63050
(314-789-2078, 586-4263)

CLEVENGER DETECTOR SALES
8206 North Oak
KANSAS CITY 64118
(816-436-0697)

THE TREASURE HUT
1315 North Main
POPLAR BLUFF 63901
(314-785-1164)

RADFORD JEWELERS
1864 South Glenstone
SPRINGFIELD 65804
(417-881-7308)

STANLEY JOHNSON CO.
2607 So. 14th
ST. JOSEPH 64503
(816-232-5163)

OZARK TREASURE CHEST
P. O. Box 417
129 Main Street
WARSAW 65355
(816-438-5445)

MONTANA

FRAN JOHNSON'S SPORT SHOP
1957 Harrison
BUTTE 59701
(406-792-3322)

ELECTRONIC PARTS
1030 S. Avenue West
P. O. Box 2126
MISSOULA 59801
(406-543-3119)

NEBRASKA

COLLINS ENTERPRISES
P. O. Box 727
BELLEVUE 68005
(402-291-1733)

EXANIMO ESTABLISHMENT
(FORMERLY SPARTAN SHOP)
335 N. William
FREMONT 68025
(402-727-9833, 721-9438)

L. P. ENTERPRISES
Box 46
SPRAGUE 68438
(402-794-5730)

NEVADA

VEGAS COIN & STAMP GALLERIES, INC.
Sahara Square Suite 28
1155 E. Sahara
LAS VEGAS 89104
(702-734-8199)

SIERRA DETECTORS
419 Flint
RENO 89501
(702-323-2712)

NEW HAMPSHIRE

DON WILSON SALES
93 So. State Street
CONCORD 03301
(603-224-5909)

STREETER ELECTRONICS
METAL DETECTOR SALES
504 Washington Street
KEENE 03431
(603-357-0229)

THE VILLAGE TRADER
U.S. Route 1
SEABROOK 03874
(603-474-2836)

NEW JERSEY

GENERAL SALES
#10 Humphrey Street
ENGLEWOOD 07631
(201-568-5563)

GEO-QUEST
78 Kenzel Avenue
NUTLEY 07110
(201-667-8170)

THE TREASURE COVE
1055 S. Clinton Avenue
TRENTON 08611
(609-393-3631, 989-7382)

NEW YORK

C-T DETECTORS
4443 Murdock Avenue
BRONX 10466
(212-325-9582)

LOST COINS ENTERPRISE
721 Mosley Road
FAIRPORT 14450
(716-223-2139)

J. PANNA'S ELECTRONIC SALES
Box 167
GENEVA 14456
(315-789-0809)

FRED BOND
2 Leech Circle So.
GLEN COVE 11542
(516-676-1380, 1310)

TRADE MART ENTERPRISES
94 Keller Avenue
KENMORE 14217
(716-875-0951)

DOC DAVE'S TREASURE FINDERS
54 Stockton Avenue, Route 206
WALTON 13856
(607-865-5188)

NORTH CAROLINA

TREASURE WORLD OF NORTH
CAROLINA
East Dixie Drive
ASHEBORO 27203
(919-629-6164)

ERNIE "CAROLINA" CURLEE
DETECTOR SALES CO.
DIVISION OF CHEMATION, INC.
3201 Cullman Avenue
CHARLOTTE 28206
(704-375-8468, 537-5115)

BARBEE DETECTOR SALES
c/o Barbee Fabrics, Inc.
P. O. Box 4235
GLEN RAVEN 27215
(919-584-7781, 584-7873)

B & R DETECTOR SALES
Route 1, Box 185-D
MONCURE 27559
(919-542-2210, 542-3832)

RUSS SIMMONS
414 Biscayne Drive
WILMINGTON 28405
(919-686-7009)

NORTH DAKOTA

CHESTER IVERSON
808 17th Avenue S.W.
MINOT 58701
(701-838-0149)

OHIO

NORTHWEST ACCESSORIES
2085 Mistyhill Drive
CINCINNATI 45240
(513-851-9164)

KILIAN DETECTOR EQUIPMENT CO.
1031 Spring Road
CLEVELAND 44109
(216-398-4779)

FOX METAL DETECTORS
RR #2, Box 312 D
LEWISBURG 45338
(513-962-2937)

KLINGLER'S ROCKS 'N THINGS
411 Bowman Road
LIMA 45804
(419-227-5294)

FORT ANCIENT TRADING
POST ANTIQUES
6 Miles East of Lebanon, Route 350
OREGONIA 45054
(513-932-3109)

WINKLE RADIO & TV
17 Mi. N. Lima Route 115
OTTAWA 45875
(419-532-3957)

STRUBLE DRUG INC. OF SHELBY
31 West Main Street
SHELBY 44875
(419-342-2136, 347-2802)

THE TREASURE CHEST
3 Mi. W. Waterville, Route 24
9204 S. River Road
WATERVILLE 43566
(419-878-6026)

OKLAHOMA

EDDIE S. FAUSETT SALES
2729 Kirby Drive
ADA 74820
(405-332-3156)

WOODROW J. RUSSEY
904 N. Juniper
JENKS 74037
(918-299-3551)

JIM PAVLU
Route 2, Box 118A
OKEENE 73763
(405-822-4810)

HOBBY WORLD
2433 Plaza Prom
Shepherd Mall
OKLAHOMA CITY 73107
(405-942-4556)

ACE'S DETECTOR SERVICE
5622 S. Pittsburg
TULSA 74135
(918-742-2214)

JOE LAWDER
R.R. #1, Box 22
TURPIN 73950
(405-854-6429)

OREGON

CARLA KAY SALVAGE CO.
471 Brule Street
COOS BAY 97420
(503-888-4015)

OREGON GOLD DREDGE LIMITED
50 Grimes Road
EUGENE 97401
(503-686-2769)

D & K DETECTOR SALES
13809 S.E. Division
PORTLAND 97236
(503-761-1521)

PENNSYLVANIA

J & D METAL LOCATING
PENN-WAY MARKET
RD #1
CONEMAUGH 15909
(814-749-9411, 322-4984)

SEALAND'S METAL DETECTORS
422 Sells Lane
GREENSBURG 15601
(412-834-3429)

SONNY'S CYCLE & SPORTING ARMS
1964 Maple Avenue, Route 213
LANGHORNE 19047
(215-752-3030)

MILLER'S TREASURE & METAL
DETECTORS
RD #1, Pettis Road
MEADVILLE 16335
(814-336-5453)

BARKER ADVERTISING
RD #5, Mitchell Rd.
NEW CASTLE 16105
(412-652-7596)

F. T. KLINGES
246 S. Sherman Street
WILKES BARRE 18702
(717-824-4277)

K. A. DETECTORS
RD 4, Box 323
WILLIAMSPORT 17701
(717-326-0867)

JACK MORGAN
MORGAN'S FURNITURE
106 N. Main Street
ZELIENOPLE 16063
(412-452-7510)

RHODE ISLAND

HOUSE OF BARGAINS
345 Warwick Avenue
WARWICK 02888
(401-781-8580)

SOUTH CAROLINA

KEN LYLES DETECTORS
122 Lazy Lane
SUMTER 29150
(803-775-8840, 775-2806)

SOUTH DAKOTA

DONCO METAL DETECTORS
2424 Canyon Lake Drive
RAPID CITY 57701
(605-343-3103)

EMPIRE DETECTOR SALES
711 S. Conklin Avenue
SIOUX FALLS 57103
(605-332-0667)

TENNESSEE

CHATTANOOGA DETECTOR SALES
3110 3rd Avenue
CHATTANOOGA 37407
(615-622-8882)

HICKORY VALLEY ELECTRIC CO.
AND METAL DETECTOR SALES
6916 Lee Hwy.
CHATTANOOGA 37421
(615-892-0525, 892-3581)

C/S DETECTOR ELECTRONICS
354 Miflin Road
Route 6
JACKSON 38301
(901-424-6319)

AMONECO
425 Madison
MEMPHIS 38103
(901-526-5054)

MID SOUTH METAL DETECTOR SALES
3179 Northwood Drive
MEMPHIS 38111
(901-452-8860)

THE COLLECTOR'S SHOP
100 Oaks Shopping Center
NASHVILLE 37204
(615-383-5996)

TEXAS

WILLIAM B. DOSS
3816 N. 11th
ABILENE 79603
(915-672-5102)

J. C. CLAXTON
2701 S. Marrs
AMARILLO 79103
(806-374-3820, 383-1613)

NILES CARTER
2103 Whitestone Drive
AUSTIN 78745
(512-444-0106)

CHESTER'S COIN SHOP
2606 International Blvd.
BROWNSVILLE 78521
(512-546-4252)

TREASURE HOUND DETECTOR SALES
400 Mitchell
BRYAN 77801
(713-779-6423, 845-2211)

BAYSIDE CERAMICS
9237 So. Padre Island Drive
CORPUS CHRISTI 78418
(512-937-1682, 937-5334)

UNITED TREASURE HUNTERS
11602 Garland Road
DALLAS 75218
(214-328-1223)

TREASURE HUNT
904 S. Crockett
DENISON 75020
(work-214-463-2110)
(home-214-463-3101)

AMERICAN CAMPING &
OUTING INDUSTRIES, INC.
P. O. Box 12564
EL PASO 79912
(915-751-7741)

B. A. STEPHENS
809 Denton Drive
EULESS 76039
(817-283-2433)

REX GROVE AUTO SUPPLY CO., INC.
4527 E. Belknap
FT. WORTH 76117
(817-838-3066, 838-9640)

ASSOCIATED TREASURE FINDERS
15006 Welcome Lane
HOUSTON 77014
(713-440-4333)

RESEARCH & RECOVERY
2803 Old Spanish Trail
HOUSTON 77054
(713-747-4647)

ALEXANDER ENTERPRISES
21 Spencer Highway
S. HOUSTON 77587
(713-946-6399)

DON GARRETT
2021 Allendale
LUFKIN 75901
(713-634-7037)

MISSION REXALL DRUG
1030 Conway
MISSION 78572
(512-585-1532)

OWENS DETECTOR SALES
5814 Kepler Drive
SAN ANTONIO 78228
(512-434-1605)

SPURGEON'S ARTIFACTS & COINS
205 W. Nueces
UVALDE 78801
(512-278-2164)

JERRY ECKHART
ECKHART'S DETECTOR SALES
3602 Armory Road
WICHITA FALLS 76302
(817-767-3939)

UTAH

BILL'S DETECTOR SALES
P.O. Box 3425
285 S. 300 E.
MT. PLEASANT 84647
(801-462-2741)

WRIGHT'S ANTIQUES
4524 Madison Avenue
OGDEN 84403
(801-479-4702)

BRYANT T. CASH
2457 West 4975 South
ROY 84067
(801-825-7858)

GALLENSON'S
220 S. State Street
SALT LAKE CITY 84111
(801-328-2016)

VIRGINIA

SUBURBAN DETECTORS
3169 Spring Street
FAIRFAX 22030
(703-273-2542)

M & R DETECTOR SALES
207 Dixie Airport Rd.
MADISON HGTS. 24572
(804-846-5036)

ESSENTIAL ELECTRONICS
10453 Medina Road
RICHMOND 23235
(804-272-5558)

H & S DETECTOR CENTER
2108 Thoroughgood Road
VIRGINIA BEACH 23455
(804-464-6072)

WASHINGTON

CACHE INN DETECTORS
17925 S.E. 313th
AUBURN 98002
(206-631-0466)

TANNER'S DIGGIN'S
4029 Boundary Trail NW
BREMERTON 98310
(206-377-2532)

THE COIN CRADLE INC.
2810 W. Kennewick Avenue Suite "E"
KENNEWICK 99336
(509-735-1507)

PEARL ELECTRONICS INC.
1300 First Avenue
SEATTLE 98101
(206-622-6200)

BOWEN'S HIDEOUT
S. 1823 Mt. Vernon
SPOKANE 99203
(509-534-4004)

WEST VIRGINIA

CHARLES MURDOCK
MURDOCK'S HOBBY SHOP
121 N. Fourth Avenue
PADEN CITY 26159
(304-337-2711)

RAY'S LEISURE TIME SHOP
P.O. Drawer E, US Highways 19 & 21
SHADY SPRINGS 25918
(304-763-3110)

WISCONSIN

PAUL'S DETECTOR SALES
1241 Bellevue Road
Bellevue Plaza
GREEN BAY 54301
(414-468-7789)

PETE'S HOBBY AND ROCK SHOP
1917-19 Winnebago Street
MADISON 53704
(608-249-2648)

WYOMING

CASPER METAL DETECTORS
SALES & RENTALS
1281 Payne And 1017 Cardiff
CASPER 82601
(307-235-6323, 234-5205)

AUSTRALIA

P. J. BRIDGE
HESPERIAN DETECTORS
P. O. Box 317
VICTORIA PARK #6100
WESTERN AUSTRALIA
(09)38-11762, 32-58575

CANADA

* CANADIAN TREASURE TRAIL
P. O. Box 22
Camden East, Ontario
CANADA KOK 1JO
(613-378-6421)

* Distributor and Service
Center for Canada

ONTARIO TREASURE SEEKERS
P. O. Box 172
Lanark, Ontario
CANADA KOA 1JO
(613-267-1080, 259-2685)

JERRY'S DETECTORS
Box 536
508-4th Avenue N.E.
Milk River, Alberta
CANADA TOK 1TO
(1-403-647-3851)

DISCOVERY DETECTORS
Box 1284
Rocky Mountain House, Alberta
CANADA TOM 1TO
(1-403-845-3718)

PIRATES COVE
3274 Danforth Avenue
Scarborough, Ontario
CANADA M1L 1C3
(1-416-691-5560)

TALL PINES TREASURE TRAIL
Box 186
Stirling, Ontario
CANADA
(613-395-2406)

L. W. ELECTRONICS
Box 42
Strathroy, Ontario
CANADA
(1-519-245-1994)

DIVERSIFIED ELECTRONICS LIMITED
1104 Franklin Street
Vancouver, British Columbia
CANADA V6A 1J6
(1-604-254-0761)

D. KEITH EDWARDS
R.R. #5
Waterford, Ontario
CANADA N0E 1Y0
(1-519-443-5193)

JOHN MENKEN
67 Darlington Street E
Yorkton, Saskatchewan
CANADA S3N OC4
(306-783-8336)

GREAT BRITAIN & EIRE

PIECES OF EIGHT
155 Robert Street
London N.W. 1
ENGLAND
(01-387-3142)

RECOMMENDED SUPPLEMENTARY BOOKS

DETECTOR OWNER'S FIELD MANUAL. Roy Lagal. Ram Publishing Company, Dallas, Texas. Nowhere else will you find the detector operating instructions that Mr. Lagal has put into this book. He shows in detail how to treasure hunt, cache hunt, prospect, search for nuggets, black sand deposits . . . in short, how to use your detector exactly as it should be used. Covers completely BFO-TR-VLF-types, P.I.'s, P.R.G.'s, P.I.P's; etc. Explains precious metals, minerals, ground conditions, and more. Fully illustrated. 236 pages. $6.95.

ELECTRONIC PROSPECTING. Charles Garrett, Bob Grant, Roy Lagal. Ram Publishing Company, Dallas, Texas. A tremendous upswing in electronic prospecting for gold and other precious metals has recently occurred. Due to high gold prices and the unlimited capabilities of the new VLF/TR metal detectors, the increased activity in prospecting has led to many fantastic discoveries. Gold is there to be found. If you have the desire and want to be successful, then this book, written by top authorities, will show you how to select (and use) from the many brands of VLF/TR's on the market those that are correctly calibrated to produce accurate metal vs. mineral identification which is so vitally necessary in the prospecting field. Illustrated. 94 pages. $3.95.

GOLD PANNING IS EASY. Roy Lagal. Ram Publishing Company, Dallas, Texas. Roy Lagal proves it! He doesn't introduce a new method; he just removes confusion surrounding old established methods. A refreshing NEW LOOK guaranteed to produce results with the "Gravity Trap" or any other pan. Special metal detector prospecting sections. This HOW, WHERE and WHEN gold panning book is a must for everyone, beginner or professional! Fully illustrated. 82 pages. $3.95.

HOW TO TEST "BEFORE BUYING" DETECTOR FIELD GUIDE. Roy Lagal. Ram Publishing Company, Dallas, Texas. Completely explains the inner workings of the BFO, TR, and discriminator types of detectors. You will learn how to test for sensitivity, stability, total response, wide scan, soil conditions, coils, Faraday shields, frequency drift, and you will be able to expose incompetent detector engineering and overly-enthusiastic, misleading advertising. If you own or are thinking of buying a detector, this book is an ABSOLUTE MUST. Fully illustrated. 64 pages. $3.95.

THE COMPLETE VLF-TR METAL DETECTOR HANDBOOK (All About Ground Canceling Metal Detectors). Roy Lagal, Charles Garrett. Ram Publishing Company, Dallas, Texas. The unparalleled capabilities of the new VLF/TR Ground Canceling metal detectors have proved these instruments to be popular. From History, Theory, and Development to Coin, Cache, and Relic Hunting, and Prospecting, these widely recognized authors have explained in detail the capabilities of the VLF/TR detectors and how they are used. Everything one needs to know about these instruments and how to use them is contained in these pages. Learn the new ground canceling detectors for the greatest possible success. Illustrated. 200 pages. $7.95.

THE JOURNALS OF EL DORADO *(Being a Descriptive Bibliography on Treasure and Subjects Pertaining Thereto. A Waybill to Discovery and Adventure).* Estee Conatser/Karl von Mueller. Ram Publishing Company, Dallas, Texas. A first-of-its-kind, this annotated list of books is devoted exclusively to books containing information of interest to treasure hunters, prospectors, and collectors of relics and bottles. It contains approximately 1,800 listings alphabetically by author and was developed as a working tool and reference for those in the treasure, small mining and prospecting fields, especially beginners. An added value is that hundreds of books are listed which do not have the word "treasure" in their titles but which contain, nevertheless, valuable treasure information. Index. 380 pages. $9.95.

TREASURE HUNTER'S MANUAL #6. Karl von Mueller. Ram Publishing Company, Dallas, Texas. Contains information to start all THers down the path to successful treasure hunting. Thousands of ideas, tips, photos, and other valuable information. 318 pages. $6.95.

TREASURE HUNTER'S MANUAL #7. Karl von Mueller. Ram Publishing Company, Dallas, Texas. The most complete, up-to-date guide to America's fastest growing hobby, written by an old master of treasure hunting. Includes topics ranging from research techniques to detector operation, from legality to gold dredging. 100% different from the THM #6. 334 pages. $6.95

SUCCESSFUL COIN HUNTING. Charles Garrett. Ram Publishing Company, Dallas, Texas. The best and most complete guide to successful coin hunting, this book explains fully the how's, where's, and when's of searching for coins and related objects. It also includes a complete explanation of how to select and use the various types of coin hunting metal detectors. Based on more than twenty years of actual in-the-field experience by the author, this volume contains a great amount of practical coin hunting information that will not be found elsewhere. Profusely illustrated with over 100 photographs. 248 pages. $5.95.

TREASURE HUNTING PAYS OFF! Charles Garrett. Ram Publishing Company, Dallas, Texas. Tells you how to begin and be successful in all phases of general treasure hunting — research; coin hunting; relic, cache, and bottle seeking; prospecting; use of metal/mineral detectors. Excellent guidebook for the beginner, plus tips and ideas for the experienced THer. Illustrated. 86 pages. $2.00.

THE COMPLETE BOOK OF COMPETITION TREASURE HUNTING. Ernie "Carolina" Curlee. Ram Publishing Company, Dallas, Texas. *NEW!* Just what the title implies! This book gives the details you need to know to sponsor or compete in an organized treasure hunt, practical information from a man who has sponsored and competed in many hunts himself. All about everything from choosing a name for a hunt and promoting it to receiving the prize you may have won. Whole sections on "How To Sponsor" and "How To Win," with a complete chapter devoted to "Understanding Your Detector." Chapter 8 features a detailed list of basic things to do and consider in planning an organized hunt. Every metal detector owner/treasure hunter can benefit from Ernie's down-to-earth, plainly written information and instructions. Many illustrations. 88 pages. $4.95.

PROFESSIONAL TREASURE HUNTER. George Mroczkowski as told to Chara Bishop. Ram Publishing Company, Dallas, Texas. *NEW!* It is a well established fact that research is 90 percent of the success of any treasure hunting endeavor. As you read this book, you will be amazed at how many treasure leads one man can come up with. You will learn how, through proper treasure hunting techniques and methods, George was able to find treasure sites, obtain permission to search (even from the U.S. Government), select and use the proper equipment, and then recover treasure in many instances. If the treasure itself was not found, extremely valuable clues and historical artifacts were located that made it all worthwhile or kept the search alive. You'll find in this book hundreds of ways how to become a better treasure hunter. Profusely illustrated. 132 pages. $6.95.